Introduction to Physical Chemistry

Introduction to Physical Chemistry

Benjamin Schultz

MURPHY & MOORE
www.murphy-moorepublishing.com

Published by Murphy & Moore Publishing,
1 Rockefeller Plaza,
New York City, NY 10020, USA

ISBN: 978-1-63987-334-0

Cataloging-in-Publication Data

Introduction to physical chemistry / Benjamin Schultz.
p. cm.
Includes bibliographical references and index.
ISBN 978-1-63987-334-0
1. Chemistry, Physical and theoretical. 2. Chemistry. I. Schultz, Benjamin.

QD453.3 .I58 2022
541--dc23

For information on all Murphy & Moore Publications
visit our website at www.murphy-moorepublishing.com

MURPHY & MOORE

TABLE OF CONTENTS

Preface VII

Chapter 1 What is Physical Chemistry? 1

- Branches of Physical Chemistry 2

Chapter 2 Chemical Kinetics 6

- Chemical Reaction 8
- Reaction Mechanism 8
- Rate Equation 11
- Arrhenius Equation 26
- Collision Theory 31
- Catalysis 37
- Enzyme Kinetics 54
- Chemical Equilibrium 83

Chapter 3 Electrochemistry 85

- Electrochemical Reaction 88
- Electrochemical Potential 101
- Electrochemical Cell 103
- Electrolysis 128
- Applications of Electrochemistry 135

Chapter 4 Chemical Thermodynamics 142

- Thermodynamic System 150
- Spontaneous Process 152
- Nonspontaneous Process 154
- The Laws of Thermodynamics 154
- Hess's Law 160
- Internal Energy 165
- Enthalpy 167

- Entropy 177
- Gibbs Free Energy 182

Chapter 5 Understanding Quantum Chemistry 194

- Electronic Structure 198
- Electronic Structure Theory 199
- Atomic Orbital 205
- Quantum Numbers 206
- Schrödinger Equation 220
- Klopman-Salem Equation 229
- Born-Oppenheimer Approximation 230

Permissions

Index

PREFACE

This book aims to help a broader range of students by exploring a wide variety of significant topics related to this discipline. It will help students in achieving a higher level of understanding of the subject and excel in their respective fields. This book would not have been possible without the unwavered support of my senior professors who took out the time to provide me feedback and help me with the process. I would also like to thank my family for their patience and support.

The study of concepts, principles and practices of atomic, subatomic, microscopic and particulate phenomena in chemical systems is known as physical chemistry. Force, statistical mechanics, energy, quantum chemistry, time and analytical dynamics are some of the concepts studied in this field. It also aims to study the effects of intermolecular forces, electrochemical cell reactions, rate of the reaction and colligative properties of materials present in a solution. The underlying concept of this field is the way in which pure physics is used to solve chemical problems. The main goal of this discipline is to predict the properties of chemical compounds and study how the atoms present in those compounds are bonded. Quantum chemistry, surface chemistry, thermochemistry and reaction kinetics are some of the disciplines related to this area of study. This book provides comprehensive insights into the field of physical chemistry. It also attempts to understand the multiple branches that fall under this discipline and how such concepts have practical applications. Those with an interest in this field would find this book helpful.

A brief overview of the book contents is provided below:

Chapter - What is Physical Chemistry?

The branch of chemistry dealing with the relations between the physical and chemical properties of substances and their chemical compositions is referred to as physical chemistry. It is divided into branches such as thermochemistry, electrochemistry, quantum chemistry, surface chemistry, etc. This is an introductory chapter which will briefly introduce all the significant aspects and branches of physical chemistry.

Chapter - Chemical Kinetics

The branch of physical chemistry which is concerned with the rates of chemical reactions is referred to as chemical kinetics. This chapter has been carefully written to provide an easy understanding of the various aspects of chemical kinetics including chemical reaction, reaction mechanism, rate equation, catalysis, enzyme kinetics, etc.

Chapter - Electrochemistry

Electrochemistry is the branch of physical chemistry which is concerned with the study of electricity and its interaction with chemical reactions. Galvanic cell, Daniell cell, primary and secondary cell are a few types of electrochemical cells used in electroplating. The topics elaborated in this chapter will help in gaining a better perspective about electrochemistry.

Chapter – Chemical Thermodynamics

The study of the interrelation of heat and work with chemical reactions or with physical changes of state within the laws of thermodynamics is known as chemical thermodynamics. Internal energy, enthalpy, entropy, Gibbs free energy, etc. are some of the concepts that fall under its domain. This chapter closely examines these key concepts of chemical thermodynamics to provide an extensive understanding of the subject.

Chapter – Understanding Quantum Chemistry

The branch of chemistry which primarily focuses on the applications of quantum mechanics in physical models and chemical systems is referred to as quantum chemistry. Quantum numbers, atomic orbitals, electronic structures, Schrödinger equation, Klopman-Salem equation, etc. are studied within it. This chapter sheds light on the subject of quantum chemistry for a thorough understanding of it.

Benjamin Schultz

What is Physical Chemistry? 1

• Branches of Physical Chemistry

The branch of chemistry dealing with the relations between the physical and chemical properties of substances and their chemical compositions is referred to as physical chemistry. It is divided into branches such as thermochemistry, electrochemistry, quantum chemistry, surface chemistry, etc. This is an introductory chapter which will briefly introduce all the significant aspects and branches of physical chemistry.

Physical chemistry is a branch of chemistry concerned with interactions and transformations of materials. Unlike other branches, it deals with the principles of physics underlying all chemical interactions (e.g., gas laws), seeking to measure, correlate, and explain the quantitative aspects of reactions. Quantum mechanics has clarified much for physical chemistry by modeling the smallest particles ordinarily dealt with in the field, atoms and molecules, enabling theoretical chemists to use computers and sophisticated mathematical techniques to understand the chemical behaviour of matter. Chemical thermodynamics deals with the relationship between heat and other forms of chemical energy, kinetics with chemical reaction rates. Subdisciplines of physical chemistry include electrochemistry, photochemistry, surface chemistry, and catalysis.

Physical Chemistry includes study of the physical properties of many different types of substances and on different scales (levels of physical detail). That is, it includes study of the following scales of chemical properties of materials:

Macroscopic

Macroscopic properties of substances describe how relatively large quantities of the substance behave as a group, e.g. melting points and boiling points, latent heats of fusion and vapourization, thermal conductivity, specific heat capacity, coefficient of linear thermal expansion, and many other physical properties.

Take the chemical substance H_2O as an example: A 'drop of water' is liquid. At lower temperatures it freezes (becoming solid 'ice') and at higher temperatures it evapourates (becoming a gas called 'steam'). The physical state of H_2O as a liquid, solid, or gas defines how closely a large quantity of individual molecules of H_2O are attracted to each other. Because the molecules are very small, there is a large quantity of H_2O molecules in a single drop of water. However, these properties do not describe the chemistry of the H_2O molecule: Two hydrogen atoms and one oxygen atom form each molecule of water, regardless of its temperature, hence if it solid ice, liquid water, or gaseous steam.

Microscopic

Microscopic properties of substances concern details of their physical properties observable only using the magnification provided by microscopes (there are, of course, different types and powers of microscopes e.g. light microscopes, electron microscopes, and more recently scanning probe microscopes). Microscopic physical properties include, for example, the shapes and structures of crystals - which can have important consequences for the behaviour of large sections of the material of which they are a part e.g. as used in bridges, aircraft, and so on.

Atomic

Atomic properties relate to elements. (Recall that elements consist of many individual atoms, whereas compounds consist of many molecules - which are, in turn, specific combinations of atoms joined together via chemical bonds).

Examples of atomic properties of elements include atomic numbers and atomic mass, e.g. the element Boron whose chemical symbol is B, has atomic number 5 and atomic mass 10.81 (strictly for a free neutral atom in ground state).

Subatomic

The study of and research into subatomic 'particles' (which are sometimes described in other ways, e.g. as 'energy' or 'waves') and their properties is often classified as part of physics, rather than chemistry. Advanced physical chemistry also includes subatomic structures and theories - which are important for nuclear chemistry.

Branches of Physical Chemistry

The different branches of physical chemistry are:

Thermodynamics

Thermodynamics is a branch of physical chemistry. It deals with the relation between heat, work, energy and temperature. Thermodynamics study dynamic nature of heat, or we can say the transfer of energy from place to place and change in its form. The main concept here is that heat is a form of energy which is equal to a specific amount of mechanical work.

There are four laws of thermodynamics. These laws define physical quantities which are energy, temperature, and entropy. This whole makes thermodynamic system at thermal equilibrium. All four laws discuss how the above mentioned physical quantities behave and work in different conditions.

Zeroth Law of Thermodynamics

If two systems are each in thermal equilibrium with a third system, they are in thermal equilibrium with each other. Zeroth law helps in defining the temperature.

First Law of Thermodynamics

When energy passes, as work, as heat, or with a matter, into or out from a system, the system's internal energy changes in accord with the law of conservation of energy. Equivalently, perpetual motion machines of the first kind (machines that produce work with no energy input) are impossible.

Second Law of Thermodynamics

In a natural thermodynamic process, the sum of the entropies of the interacting thermodynamic systems increases. Equivalently, perpetual motion machines of the second kind (machines that spontaneously convert thermal energy into mechanical work) are impossible.

Third Law of Thermodynamics

The entropy of a system approaches a constant value as the temperature approaches absolute zero. With the exception of non-crystalline solids (glasses), the entropy of a system at absolute zero is typically close to zero.

Chemical Kinetics

The second branch of physical chemistry is Chemical kinetics. This branch of physical Chemistry also known as Reaction kinetics, studies about the rate of chemical processes. This branch of physical chemistry investigates the influence of various experimental conditions on a chemical reaction. This helps to find out chemical mechanism and transition states. It also helps in defining the characteristics of the chemical reaction.

Photochemistry

This branch of physical chemistry studies the chemical effects of light. Usually, the term Photochemistry defines a chemical reaction occurred due to absorption of UV, Visible light or Infrared radiations.

Photochemistry has a vital role because it forms the basis of photosynthesis, the formation of vitamin D and vision. Photochemical reactions are different from other reactions.

Photochemical reactions create high energy that is not possible to generate with heat. This makes to cross large activation barriers in the least time.

Quantum Chemistry

This is a branch of Physical chemistry that focus on the applications of quantum mechanics in various physical models and different experiments of chemistry. This branch is also famous with the name quantum mechanics.

Different Models in quantum mechanics:

- Wave model.
- Valence model.
- Molecular model.

Electrochemistry

This branch of physical chemistry studies the movement of electrons due to various chemical processes. Movement of electrons is known as electricity. Electricity is generated from the displacement of electrons from one element to another in a reaction known as a redox reaction. This reaction is also famous with the name oxidation-reduction reaction. This reaction involves the change of state of single or more elements. On losing an electron by a substance, oxidation state increases, and it is called as oxidized. On gaining the electron by a substance, its oxidation state gets reduce.

Material Chemistry

This branch of physical chemistry also sometimes known as solid state chemistry. It is a study of structure, synthesis and different property of solid materials. This branch has overlap with thermodynamics, mineralogy, crystallography, metallurgy, ceramics, and electronics with a synthesis of unique materials and their features. Solids are studied under two forms one is crystalline and other is amorphous. This depends on the arrangement of constituent particles.

Biophysical Chemistry

This branch of physical chemistry involves the concepts of physics and chemistry together to research on biological systems. In this research, scientists seek information on the various phenomenon in biological systems. These systems can be molecules or supra molecules of a bio system.

Material Science

This field is also popular with the name material science and engineering. This field is related to the discovery and designing of new materials which are specifically solid in nature. The origin of material science was back in the age of enlightenment. During this period researchers and scientists utilized their analytical knowledge of chemistry, engineering and physics to understand the old age metallurgy and mineralogy.

Spectroscopy

Spectroscopy is also a branch of physical chemistry. It deals with the interaction of electromagnetic radiations with the matter. This branch came into existence with the study of dispersion of white light through a glass prism according to the wavelength. After a few years, researchers also started studying radioactive energy interaction in the electromagnetic spectrum.

The term spectrograph or spectroscopy are used to indicate a measurement of radiation intensity and also helps in experimental spectroscopic methods. Here is the list of some spectral measurement devices:

- Spectrophotometers.
- Spectrographs.
- Spectral analyzers.
- Spectrometers.

Physical Organic Chemistry

The name was first suggested by Louis Hammett in the year 1940. This branch of physical chemistry is closely associated with organic chemistry. It focuses on the relationship between chemical structures and reactivity. Here in this branch of physical chemistry scientists use tools of physical chemistry. These tools allow to study different organic molecules. The research has some focus points which are relative chemical stabilities, reactive intermediates, transition states, starting materials, a product of chemical reactions non covalent aspects of solvation and molecular interactions. These aspects influence chemical reactivity. Studies of these frameworks indicate how changes of structure in liquid or solid states make an impact on the reaction mechanism.

Chemical Kinetics 2

- Chemical Reaction
- Reaction Mechanism
- Rate Equation
- Arrhenius Equation
- Collision Theory
- Catalysis
- Enzyme Kinetics
- Chemical Equilibrium

The branch of physical chemistry which is concerned with the rates of chemical reactions is referred to as chemical kinetics. This chapter has been carefully written to provide an easy understanding of the various aspects of chemical kinetics including chemical reaction, reaction mechanism, rate equation, catalysis, enzyme kinetics, etc.

Chemical kinetics is the study of chemical processes and rates of reactions. This includes the analysis of conditions that affect speed of a chemical reaction, understanding reaction mechanisms and transition states, and forming mathematical models to predict and describe a chemical reaction. The rate of a chemical reaction usually has units of sec^{-1}, however, kinetics experiments may span several minutes, hours, or even days.

The field of chemical kinetics developed from the law of mass action, formulated in 1864 by Peter Waage and Cato Guldberg. The law of mass action states the speed of a chemical reaction is proportional to the amount of reactants. Jacobus van't Hoff studied chemical dynamics. His 1884 publication "Etudes de dynamique chimique" led to the 1901 Nobel Prize in Chemistry (which was the first year the Nobel prize was awarded). Some chemical reactions may involve complicated kinetics, but the basic principles of kinetics are learned in high school and college general chemistry classes.

Rate Laws and Rate Constants

Experimental data is used to find reaction rates, from which rate laws and chemical kinetics rate constants are derived by applying the law of mass action. Rate laws allow for simple calculations for zero order reactions, first order reactions, and second order reactions.

- The rate of a zero-order reaction is constant and independent of the concentration of reactants.rate = k.
- The rate of a first-order reaction is proportional to the concentration of one reactants:rate = k[A].
- The rate of a second order reaction has a rate proportional to the square of the concentration of a single reactant or else the product of the concentration of two reactants. rate = $k[A]^2$ or k[A][B].

Rate laws for individual steps must be combined to derive laws for more complex chemical reactions. For these reactions:

- There is a rate-determining step that limits the kinetics.
- The Arrhenius equation and Eyring equations may be used to experimentally determine activation energy.
- Steady-state approximations may be applied to simplify the rate law.

Factors that affect Chemical Reaction Rate

Chemical kinetics predicts the rate of a chemical reaction will be increased by factors that increase the kinetic energy of the reactants (up to a point), leading to increased likelihood the reactants will interact with each other. Similarly, factors that decrease the chance of reactants colliding with each other may be expected to lower the reaction rate. The main factors that affect reaction rate are:

- Concentration of reactants (increasing concentration increases reaction rate).
- Temperature (increasing temperature increases reaction rate, up to a point).
- Presence of catalysts (catalysts offer a reaction a mechanism that requires a lower activation energy, so the presence of a catalyst increases the rate of a reaction).
- Physical state of reactants (reactants in the same phase may come into contact via thermal action, but surface area and agitation affect reactions between reactants in different phases).
- Pressure (for reactions involving gases, raising pressure increases the collisions between reactants, increasing reaction rate).

Note that while chemical kinetics can predict the rate of a chemical reaction, it does not determine the extent to which the reaction occurs. Thermodynamics is used to predict equilibrium.

Chemical Reaction

Chemical reaction is a process in which one or more substances, the reactants, are converted to one or more different substances, the products. Substances are either chemical elements or compounds. A chemical reaction rearranges the constituent atoms of the reactants to create different substances as products.

Chemical reactions are an integral part of technology, of culture, and indeed of life itself. Burning fuels, smelting iron, making glass and pottery, brewing beer, and making wine and cheese are among many examples of activities incorporating chemical reactions that have been known and used for thousands of years. Chemical reactions abound in the geology of Earth, in the atmosphere and oceans, and in a vast array of complicated processes that occur in all living systems.

Chemical reactions must be distinguished from physical changes. Physical changes include changes of state, such as ice melting to water and water evaporating to vapour. If a physical change occurs, the physical properties of a substance will change, but its chemical identity will remain the same. No matter what its physical state, water (H_2O) is the same compound, with each molecule composed of two atoms of hydrogen and one atom of oxygen. However, if water, as ice, liquid, or vapour, encounters sodium metal (Na), the atoms will be redistributed to give the new substances molecular hydrogen (H_2) and sodium hydroxide (NaOH). By this, we know that a chemical change or reaction has occurred.

Reaction Mechanism

In chemistry, a mechanism of reaction is the step by step sequence of elementary reactions by which overall chemical change occurs.

A chemical mechanism is a theoretical conjecture that tries to describe in detail what takes place at each stage of an overall chemical reaction. The detailed steps of a reaction are not observable in most cases. The conjectured mechanism is chosen because it is thermodynamically feasible, and has experimental support in isolated intermediates or other quantitative and qualitative characteristics of the reaction. It also describes each reactive intermediate, activated complex, and transition state, and which bonds are broken (and in what order), and which bonds are formed (and in what order). A complete mechanism must also explain the reason for the reactants and catalyst used, the stereochemistry observed in reactants and products, all products formed and the amount of each.

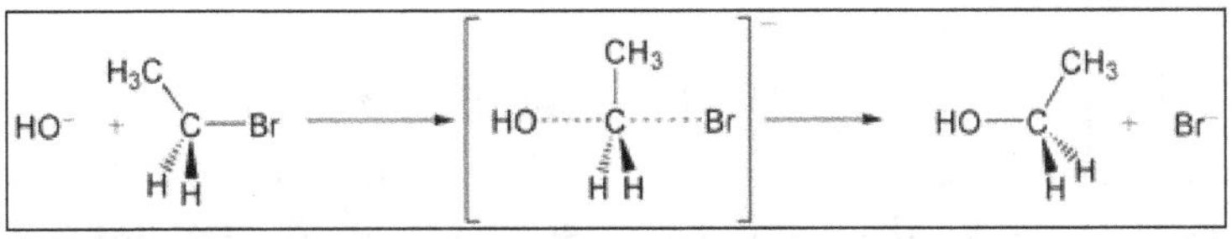

S_N2 reaction mechanism. Note the negatively charged transition state in brackets in which the central carbon atom in question shows five bonds, an unstable condition.

The electron or arrow pushing method is often used in illustrating a reaction mechanism.

A reaction mechanism must also account for the order in which molecules react. Often what appears to be a single-step conversion is in fact a multistep reaction.

Reaction Intermediates

Reaction intermediates are chemical species, often unstable and short-lived (however sometimes can be isolated), which are not reactants or products of the overall chemical reaction, but are temporary products and reactants in the mechanism's reaction steps. Reaction intermediates are often free radicals or ions.

The kinetics (relative rates of the reaction steps and the rate equation for the overall reaction) are explained in terms of the energy needed for the conversion of the reactants to the proposed transition states (molecular states that corresponds to maxima on the reaction coordinates, and to saddle points on the potential energy surface for the reaction).

Determination of Reaction Mechanism

Information about the mechanism of a reaction is often provided by the use of chemical kinetics to determine the rate equation and the reaction order in each reactant.

Consider the following reaction for example:

$$CO + NO_2 \rightarrow CO_2 + NO$$

In this case, experiments have determined that this reaction takes place according to the rate law $r = k[NO_2]^2$. This form suggests that the rate-determining step is a reaction between two molecules of NO_2. A possible mechanism for the overall reaction that explains the rate law is:

$$2\,NO_2 \rightarrow NO_3 + NO\ (\text{slow})$$

$$NO_3 + CO \rightarrow NO_2 + CO_2\ (\text{fast})$$

Each step is called an elementary step, and each has its own rate law and molecularity. The elementary steps should add up to the original reaction. (Meaning, if we were to cancel out all the molecules that appear on both sides of the reaction, we would be left with the original reaction).

When determining the overall rate law for a reaction, the slowest step is the step that determines the reaction rate. Because the first step (in the above reaction) is the slowest step, it is the rate-determining step. Because it involves the collision of two NO_2 molecules, it is a bimolecular reaction with a rate law of $r = k[NO_2]^2$.

Other reactions may have mechanisms of several consecutive steps. In organic chemistry, the reaction mechanism for the benzoin condensation, put forward in 1903 by A. J. Lapworth, was one of the first proposed reaction mechanisms.

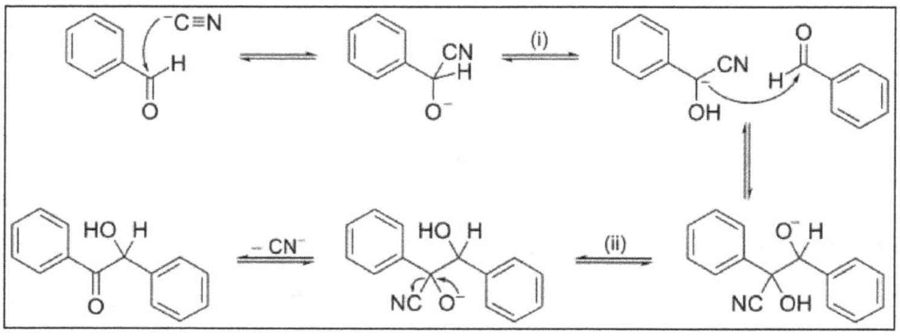

Benzoin condensation reaction mechanism. Cyanide ion (CN^-) acts as a catalyst here, entering at the first step and leaving in the last step. Proton (H^+) transfers occur at (i) and (ii). The arrow pushing method is used in some of the steps to show where electron pairs go.

A chain reaction is an example of a complex mechanism, in which the propagation steps form a closed cycle.

Other Experimental Methods to Determine Mechanism

Many experiments that suggest the possible sequence of steps in a reaction mechanism have been designed, including:

- Measurement of the effect of temperature (Arrhenius equation) to determine the activation energy.
- Spectroscopic observation of reaction intermediates.
- Determination of the stereochemistry of products, for example in nucleophilic substitution reactions.
- Measurement of the effect of isotopic substitution on the reaction rate.
- for reactions in solution, measurement of the effect of pressure on the reaction rate to determine the volume change on formation of the activated complex.

- For reactions of ions in solution, measurement of the effect of ionic strength on the reaction rate.
- Direct observation of the activated complex by pump-probe spectroscopy.
- Infrared chemiluminescence to detect vibrational excitation in the products
- Electrospray ionization mass spectrometry.
- Crossover experiments.

Theoretical Modeling

A correct reaction mechanism is an important part of accurate predictive modeling. For many combustion and plasma systems, detailed mechanisms are not available or require development.

Even when information is available, identifying and assembling the relevant data from a variety of sources, reconciling discrepant values and extrapolating to different conditions can be a difficult process without expert help. Rate constants or thermochemical data are often not available in the literature, so computational chemistry techniques or group additivity methods must be used to obtain the required parameters.

Computational chemistry methods can also be used to calculate potential energy surfaces for reactions and determine probable mechanisms.

Molecularity

Molecularity in chemistry is the number of colliding molecular entities that are involved in a single reaction step.

- A reaction step involving one molecular entity is called unimolecular.
- A reaction step involving two molecular entities is called bimolecular.
- A reaction step involving three molecular entities is called trimolecular or termolecular.

In general, reaction steps involving more than three molecular entities do not occur, because is statistically improbable in terms of Maxwell distribution to find such transition state.

Rate Equation

The rate law or rate equation for a chemical reaction is an equation that links the reaction rate with the concentrations or pressures of the reactants and constant parameters (normally rate coefficients and partial reaction orders). For many reactions the rate is given by a power law such as:

$$v = k[\mathrm{A}]^x[\mathrm{B}]^y$$

where [A] and [B] express the concentration of the species A and B (usually in moles per liter (molarity, M)). The exponents x and y are the partial *orders of reaction* for A and B and the *overall* reaction order is the sum of the exponents. These are often positive integers, but they may also be zero, fractional, or negative. The constant k is the reaction rate constant or *rate coefficient* of the reaction. Its value may depend on conditions such as temperature, ionic strength, surface area of an adsorbent, or light irradiation.

Elementary (single-step) reactions and reaction steps have reaction orders equal to the stoichiometric coefficients for each reactant. The overall reaction order, i.e. the sum of stoichiometric coefficients of reactants, is always equal to the molecularity of the elementary reaction. However complex (multi-step) reactions may or may not have reaction orders equal to their stoichiometric coefficients. This implies that the order and the rate equation of a given reaction cannot be reliably deduced from the stoichiometry and must be determined experimentally, since an unknown reaction mechanism could be either elementary or complex. When the experimental rate equation has been determined, it is often of use for deduction of the reaction mechanism.

The rate equation of a reaction with an assumed multi-step mechanism can often be derived theoretically using quasi-steady state assumptions from the underlying elementary reactions, and compared with the experimental rate equation as a test of the assumed mechanism. The equation may involve a fractional order, and may depend on the concentration of an intermediate species.

A reaction can also have an *undefined* reaction order with respect to a reactant if the rate is not simply proportional to some power of the concentration of that reactant; for example, one cannot talk about reaction order in the rate equation for a bimolecular reaction between adsorbed molecules:

$$v = k\frac{K_1K_2C_AC_B}{(1+K_1C_A+K_2C_B)^2}.$$

Consider a typical chemical reaction in which two reactants A and B combine to form a product C:

$$\mathrm{A} + 2\mathrm{B} \rightarrow 3\mathrm{C}.$$

This can also be written

$$0 = -\mathrm{A} - 2\mathrm{B} + 3\mathrm{C}.$$

The prefactors -1, -2 and 3 (with negative signs for reactants because they are consumed) are known as stoichiometric coefficients. One molecule of A combines with two of B to form 3 of C, so if we use the symbol [X] for the number of moles of chemical X,

$$-\frac{d[\mathrm{A}]}{dt}=-\frac{1}{2}\frac{d[\mathrm{B}]}{dt}=\frac{1}{3}\frac{d[\mathrm{C}]}{dt}.$$

If the reaction takes place in a closed system at constant temperature and volume, without a build-up of reaction intermediates, the *reaction rate r* is defined as

$$v=\frac{1}{v_i}\frac{d[\mathrm{X}_i]}{dt},$$

where v_i is the stoichiometric coefficient for chemical X_i.

The reaction rate has some functional dependence on the concentrations of the reactants,

$$v=f([\mathrm{A}],[\mathrm{B}],\ldots),$$

and this dependence is known as the *rate equation* or *rate law*. This law generally cannot be deduced from the chemical equation and must be determined by experiment.

Power Laws

A common form for the rate equation is a power law:

$$v=k[A]^x[B]^y\ldots$$

The constant *k* is called the *rate constant*. The exponents, which can be fractional, are called *partial orders of reaction* and their sum is the overall order of reaction.

In a dilute solution, an elementary reaction (one having a single step with a single transition state) is empirically found to obey the law of mass action. This predicts that the rate depends only on the concentrations of the reactants, raised to the powers of their stoichiometric coefficients.

Determination of Reaction Order

Method of Initial Rates

The natural logarithm of the power-law rate equation is

$$\ln v=\ln k+x\ln[A]+y\ln[B]+\ldots$$

This can be used to estimate the order of reaction of each reactant. For example, the initial rate can be measured in a series of experiments at different initial concentrations of reactant A with all other concentrations [B], [C], ... kept constant, so that,

$$v=x\ln[A]+\text{constant}.$$

The slope of a graph of ln v as a function of ln[A] then corresponds to the order x with respect to reactant A.

However, this method is not always reliable because

1. Measurement of the initial rate requires accurate determination of small changes in concentration in short times (compared to the reaction half-life) and is sensitive to errors.
2. The rate equation will not be completely determined if the rate also depends on substances not present at the beginning of the reaction, such as intermediates or products.

Integral Method

The tentative rate equation determined by the method of initial rates is therefore normally verified by comparing the concentrations measured over a longer time (several half-lives) with the integrated form of the rate equation.

For example, the integrated rate law for a first-order reaction is

$$\ln[A] = -kt + \ln[A]_0,$$

where [A] is the concentration at time t and $[A]_0$ is the initial concentration at zero time. The first-order rate law is confirmed if ln[A] is in fact a linear function of time. In this case the rate constant k is equal to the slope with sign reversed.

Method of Flooding

The partial order with respect to a given reactant can be evaluated by the method of flooding (or of isolation) of Ostwald. In this method, the concentration of one reactant is measured with all other reactants in large excess so that their concentration remains essentially constant. For a reaction $a{\cdot}A\ +\ b{\cdot}B\ \rightarrow\ c{\cdot}C$ with rate law: $v = k\cdot[\mathrm{A}]^x\cdot[\mathrm{B}]^y$, the partial order x with respect to A is determined using a large excess of B. In this case $v = k'\cdot[\mathrm{A}]^x$ with $k' = k\cdot[\mathrm{B}]^y$, and x may be determined by the integral method. The order y with respect to B under the same conditions (with B in excess) is determined by a series of similar experiments with a range of initial concentration $[\mathrm{B}]_0$ so that the variation of k' can be measured.

Zero Order

For zero-order reactions, the reaction rate is independent of the concentration of a reactant, so that changing its concentration has no effect on the speed of the reaction. Thus, the concentration changes linearly with time. This may occur when there is a bottleneck which limits the number of reactant molecules that can react at the same time, for example if the reaction requires contact with an enzyme or a catalytic surface.

Many enzyme-catalyzed reactions are zero order, provided that the reactant concentration is much greater than the enzyme concentration which controls the rate, so that the enzyme is *saturated*. For example, the biological oxidation of ethanol to acetaldehyde by the enzyme liver alcohol dehydrogenase (LADH) is zero order in ethanol.

Similarly reactions with heterogeneous catalysis can be zero order if the catalytic surface is saturated. For example, the decomposition of phosphine (PH_3) on a hot tungsten surface at high pressure is zero order in phosphine which decomposes at a constant rate.

In homogeneous catalysis zero order behavior can come about from reversible inhibition. For example, ring-opening metathesis polymerization using third-generation Grubbs catalyst exhibits zero order behavior in catalyst due to the reversible inhibition that is occur between the pyridine and the ruthenium center.

First Order

A first order reaction depends on the concentration of only one reactant (a unimolecular reaction). Other reactants can be present, but each will be zero order. The rate law for such a reaction is,

$$-\frac{d[A]}{dt}=k[A],$$

The half-life is independent of the starting concentration and is given by $t_{1/2}=\frac{\ln(2)}{k}$.

Examples of such reactions are:

$$H_2O_2(l) \rightarrow H_2O(l)+\frac{1}{2}O_2(g)$$

$$SO_2Cl_2(l) \rightarrow SO_2(g)+Cl_2(g)$$

$$2\,N_2O_5(g) \rightarrow 4\,NO_2(g)+O_2(g)$$

In organic chemistry, the class of S_N1 (nucleophilic substitution unimolecular) reactions consists of first-order reactions. For example, in the reaction of aryldiazonium ions with nucleophiles in aqueous solution $ArN_2^+ + X^- \rightarrow ArX + N_2$, the rate equation is $v=k[ArN_2^+]$, where Ar indicates an aryl group.

Second Order

A reaction is said to be second order when the overall order is two. The rate of a second-order reaction may be proportional to one concentration squared $v=k[A]^2$, or (more commonly) to the product of two concentrations $v=k[A][B]$. As an example of the first type, the reaction $NO_2+CO \rightarrow NO+CO_2$ is second-order in the reactant NO_2

and zero order in the reactant CO. The observed rate is given by $v = k[NO_2]^2$, and is independent of the concentration of CO.

For the rate proportional to a single concentration squared, the time dependence of the concentration is given by:

$$\frac{1}{[A]} = \frac{1}{[A]_0} + kt$$

The time dependence for a rate proportional to two unequal concentrations is,

$$\frac{[A]}{[B]} = \frac{[A]_0}{[B]_0} e^{([A]_0 - [B]_0)kt};$$

if the concentrations are equal, they satisfy the previous equation.

The second type includes nucleophillic addition-elimination reactions, such as the alkaline hydrolysis of ethyl acetate:

$$CH_3COOC_2H_5 + OH^- \rightarrow CH_3COO^- + C_2H_5OH$$

This reaction is first-order in each reactant and second-order overall: $r = k[CH_3COOC_2H_5][OH^-]$

If the same hydrolysis reaction is catalyzed by imidazole, the rate equation becomes $v = k[\text{imidazole}][CH_3COOC_2H_5]$. The rate is first-order in one reactant (ethyl acetate), and also first-order in imidazole which as a catalyst does not appear in the overall chemical equation.

Another well-known class of second-order reactions are the S_N2 (bimolecular nucleophilic substitution) reactions, such as the reaction of n-butyl bromide with sodium iodide in acetone:

$$CH_3CH_2CH_2CH_2Br + NaI \rightarrow CH_3CH_2CH_2CH_2I + NaBr \downarrow$$

This same compound can be made to undergo a bimolecular (E2) elimination reaction, another common type of second-order reaction, if the sodium iodide and acetone are replaced with sodium tert-butoxide as the salt and tert-butanol as the solvent:

$$CH_3CH_2CH_2CH_2Br + NaOt\text{-}Bu \rightarrow CH_3CH_2CH = CH_2 + NaBr + HOt\text{-}Bu$$

Pseudo-first Order

If the concentration of a reactant remains constant (because it is a catalyst, or because it is in great excess with respect to the other reactants), its concentration can be included in the rate constant, obtaining a *pseudo–first-order* (or occasionally pseudo–second-order) rate equation. For a typical second-order reaction with rate equation v = k[A][B], if the concentration of reactant B is constant then v = k[A][B] = k'[A], where the pseudo–first-order rate constant k' = k[B]. The second-order rate equation has been reduced to a pseudo–first-order rate equation, which makes the treatment to obtain an integrated rate equation much easier.

One way to obtain a pseudo-first order reaction is to use a large excess of one reactant (say, [B]≫[A]) so that, as the reaction progresses, only a small fraction of the reactant in excess (B) is consumed, and its concentration can be considered to stay constant. For example, the hydrolysis of esters by dilute mineral acids follows pseudo-first order kinetics where the concentration of water is present in large excess:

$$CH_3COOCH_3 + H_2O \rightarrow CH_3COOH + CH_3OH$$

The hydrolysis of sucrose in acid solution is often cited as a first-order reaction with rate r = k[sucrose]. The true rate equation is third-order, r = k[sucrose][H^+][H_2O]; however, the concentrations of both the catalyst H^+ and the solvent H_2O are normally constant, so that the reaction is pseudo–first-order.

Fractional Order

In fractional order reactions, the order is a non-integer, which often indicates a chemical chain reaction or other complex reaction mechanism. For example, the pyrolysis of acetaldehyde (CH_3CHO) into methane and carbon monoxide proceeds with an order of 1.5 with respect to acetaldehyde: $r = k[CH_3CHO]^{3/2}$. The decomposition of phosgene ($COCl_2$) to carbon monoxide and chlorine has order 1 with respect to phosgene itself and order 0.5 with respect to chlorine: $v = k[COCl_2]\,[Cl_2]^{1/2}$.

The order of a chain reaction can be rationalized using the steady state approximation for the concentration of reactive intermediates such as free radicals. For the pyrolysis of acetaldehyde, the Rice-Herzfeld mechanism is:

Initiation:

$$CH_3CHO \rightarrow \bullet CH_3 + \bullet CHO$$

Propagation:

$$\bullet CH_3 + CH_3CHO \rightarrow CH_3CO\bullet + CH_4$$

$$CH_3CO\bullet \rightarrow \bullet CH_3 + CO$$

Termination:

$$2\ \bullet CH_3 \rightarrow C_2H_6$$

where • denotes a free radical. To simplify the theory, the reactions of the •CHO to form a second $\bullet CH_3$ are ignored.

In the steady state, the rates of formation and destruction of methyl radicals are equal, so that,

$$\frac{d[.\mathrm{CH_3}]}{dt} = k_i[\mathrm{CH_3CHO}] - k_t[.\mathrm{CH_3}]^2 = 0,$$

so that the concentration of methyl radical satisfies,

$$[.\mathrm{CH_3}] \;\propto\; [\mathrm{CH_3CHO}]^{\frac{1}{2}}.$$

The reaction rate equals the rate of the propagation steps which form the main reaction products CH_4 and CO:

$$v = \frac{d[CH_4]}{dt} = k_p[.CH_3][CH_3CHO] \;\propto\; [CH_3CHO]^{\frac{3}{2}}$$

in agreement with the experimental order of 3/2.

Complex Laws

Mixed Order

More complex rate laws have been described as being *mixed order* if they approximate to the laws for more than one order at different concentrations of the chemical species involved. For example, a rate law of the form $v = k_1[A] + k_2[A]^2$ represents concurrent first order and second order reactions (or more often concurrent pseudo-first order and second order) reactions, and can be described as mixed first and second order. For sufficiently large values of [A] such a reaction will approximate second order kinetics, but for smaller [A] the kinetics will approximate first order (or pseudo-first order). As the reaction progresses, the reaction can change from second order to first order as reactant is consumed.

Another type of mixed-order rate law has a denominator of two or more terms, often because the identity of the rate-determining step depends on the values of the concentrations. An example is the oxidation of an alcohol to a ketone by hexacyanoferrate (III) ion $[Fe(CN)_6^{3-}]$ with ruthenate (VI) ion (RuO_4^{2-}) as catalyst. For this reaction, the rate of disappearance of hexacyanoferrate (III) is $v = \dfrac{[Fe(CN)_6]^{2-}}{k_\alpha + k_\beta[Fe(CN)_6]^{2-}}$

This is zero-order with respect to hexacyanoferrate (III) at the onset of the reaction (when its concentration is high and the ruthenium catalyst is quickly regenerated), but changes to first-order when its concentration decreases and the regeneration of catalyst becomes rate-determining.

Notable mechanisms with mixed-order rate laws with two-term denominators include:

- Michaelis-Menten kinetics for enzyme-catalysis: first-order in substrate (second-order overall) at low substrate concentrations, zero order in substrate (first-order overall) at higher substrate concentrations.
- The Lindemann mechanism for unimolecular reactions: second-order at low pressures, first-order at high pressures.

Negative Order

A reaction rate can have a negative partial order with respect to a substance. For example, the conversion of ozone (O_3) to oxygen follows the rate equation $v = k\frac{[O_3]^2}{[O_2]}$ in an excess of oxygen. This corresponds to second order in ozone and order (-1) with respect to oxygen.

When a partial order is negative, the overall order is usually considered as undefined. In the above example for instance, the reaction is not described as first order even though the sum of the partial orders is $2+(-1)=1$, because the rate equation is more complex than that of a simple first-order reaction.

Opposed Reactions

A pair of forward and reverse reactions may occur simultaneously with comparable speeds. For example, A and B react into X and Y and vice versa (*s, t, u,* and *w* are the stoichiometric coefficients):

$$sA + tB \leftrightarrow uX + wY$$

The reaction rate expression for the above reactions (assuming each one is elementary) can be expressed as:

$$v = k_1[A]^s[B]^t - k_2[X]^u[Y]^w$$

where: k_1 is the rate coefficient for the reaction that consumes A and B; k_2 is the rate coefficient for the backwards reaction, which consumes X and Y and produces A and B.

The constants k_1 and k_2 are related to the equilibrium coefficient for the reaction (K) by the following relationship (set v=0 in balance):

$$k_1[A]^s[B]^t = k_2[X]^u[Y]^w$$

$$K = \frac{[X]^u[Y]^w}{[A]^s[B]^t} = \frac{k_1}{k_2}$$

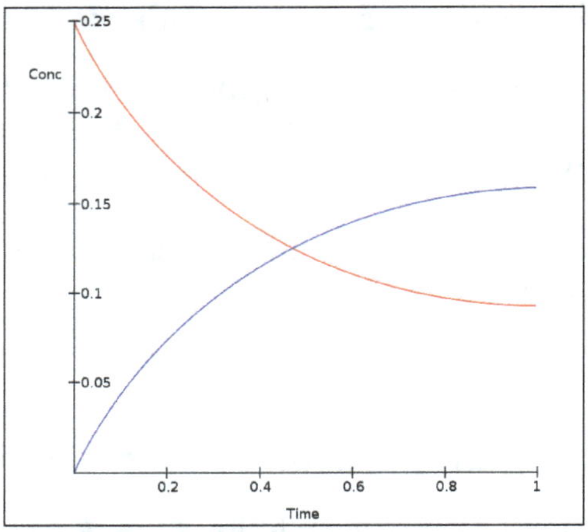

Concentration of A (A_0 = 0.25 mole/l) and B versus time reaching equilibrium k_f = 2 min^{-1} and k_r = 1 min^{-1}.

Simple Example:

In a simple equilibrium between two species:

$$A \rightleftharpoons B$$

Where the reactions starts with an initial concentration of A, $[A]_0$, with an initial concentration of 0 for B at time t=0.

Then the constant K at equilibrium is expressed as:

$$K \stackrel{\text{def}}{=} \frac{k_f}{k_b} = \frac{[B]_e}{[A]_e}$$

Where $[A]_e$ and $[B]_e$ are the concentrations of A and B at equilibrium, respectively.

The concentration of A at time t, $[A]_t$, is related to the concentration of B at time t, $[B]_t$, by the equilibrium reaction equation:

$$[A]_t = [A]_0 - [B]_t$$

The term $[B]_0$ is not present because, in this simple example, the initial concentration of B is 0.

This applies even when time t is at infinity; i.e., equilibrium has been reached:

$$[A]_e = [A]_0 - [B]_e$$

then it follows, by the definition of K, that,

$$[B]_e = x = \frac{k_f}{k_f + k_b}[A]_0$$

and, therefore,

$$[A]_e = [A]_0 - x = \frac{k_b}{k_f + k_b}[A]_0$$

These equations allow us to uncouple the system of differential equations, and allow us to solve for the concentration of A alone.

The reaction equation, given previously as:

$$v = k_1[A]^s[B]^t - k_2[X]^u[Y]^v$$

$$-\frac{d[A]}{dt} = k_f[A]_t - k_b[B]_t$$

The derivative is negative because this is the rate of the reaction going from A to B, and therefore the concentration of A is decreasing. To simplify annotation, let x be $[A]_t$, the concentration of A at time t. Let x_e be the concentration of A at equilibrium. Then:

$$-\frac{d[A]}{dt} = k_f[A]_t - k_b[B]_t$$

$$-\frac{dx}{dt} = k_f x - k_b[B]_t$$

$$= k_f x - k_b([A]_0 - x)$$

$$= (k_f + k_b)x - k_b[A]_0$$

Since:

$$k_f + k_b = k_b\frac{[A]_0}{x_e}$$

The reaction rate becomes:

$$\frac{dx}{dt} = \frac{k_b[A]_0}{x_e}(x_e - x)$$

which results in:

$$\ln\left(\frac{[A]_0 - [A]_e}{[A]_t - [A]_e}\right) = (k_f + k_b)t$$

A plot of the negative natural logarithm of the concentration of A in time minus the concentration at equilibrium versus time t gives a straight line with slope $k_f + k_b$. By measurement of A_e and B_e the values of *K* and the two reaction rate constants will be known.

Generalization of Simple Example

If the concentration at the time $t = 0$ is different from above, the simplifications above are invalid, and a system of differential equations must be solved. However, this system can also be solved exactly to yield the following generalized expressions:

$$[A]=[A]_0\frac{1}{k_f+k_b}\left(k_b+k_f e^{-(k_f+k_b)t}\right)+[B]_0\frac{k_b}{k_f+k_b}\left(1-e^{-(k_f+k_b)t}\right)$$

$$[B]=[A]_0\frac{k_f}{k_f+k_b}\left(1-e^{-(k_f+k_b)t}\right)+[B]_0\frac{1}{k_f+k_b}\left(k_f+k_b e^{-(k_f+k_b)t}\right)$$

When the equilibrium constant is close to unity and the reaction rates very fast for instance in conformational analysis of molecules, other methods are required for the determination of rate constants for instance by complete lineshape analysis in NMR spectroscopy.

Consecutive Reactions

If the rate constants for the following reaction are k_1 and k_2; $A \rightarrow B \rightarrow C$, then the rate equation is:

For reactant A: $\frac{d[A]}{dt}=-k_1[A]$

For reactant B: $\frac{d[B]}{dt}=k_1[A]-k_2[B]$

For product C: $\frac{d[C]}{dt}=k_2[B]$

With the individual concentrations scaled by the total population of reactants to become probabilities, linear systems of differential equations such as these can be formulated as a master equation. The differential equations can be solved analytically and the integrated rate equations are:

$$[A]=[A]_0 e^{-k_1 t}$$

$$[B]=\begin{cases}[A]_0\frac{k_1}{k_2-k_1}\left(e^{-k_1 t}-e^{-k_2 t}\right)+[B]_0 e^{-k_2 t} & k_1 \neq k_2\\ [A]_0 k_1 t e^{-k_1 t}+[B]_0 e^{-k_1 t} & \text{otherwise}\end{cases}$$

$$[C]=\begin{cases}[A]_0\left(1+\frac{k_1 e^{-k_2 t}-k_2 e^{-k_1 t}}{k_2-k_1}\right)+[B]_0\left(1-e^{-k_2 t}\right)+[C]_0 & k_1 \neq k_2\\ [A]_0\left(1-e^{-k_1 t}-k_1 t e^{-k_1 t}\right)+[B]_0\left(1-e^{-k_1 t}\right)+[C]_0 & \text{otherwise}\end{cases}$$

The steady state approximation leads to very similar results in an easier way.

Parallel or Competitive Reactions

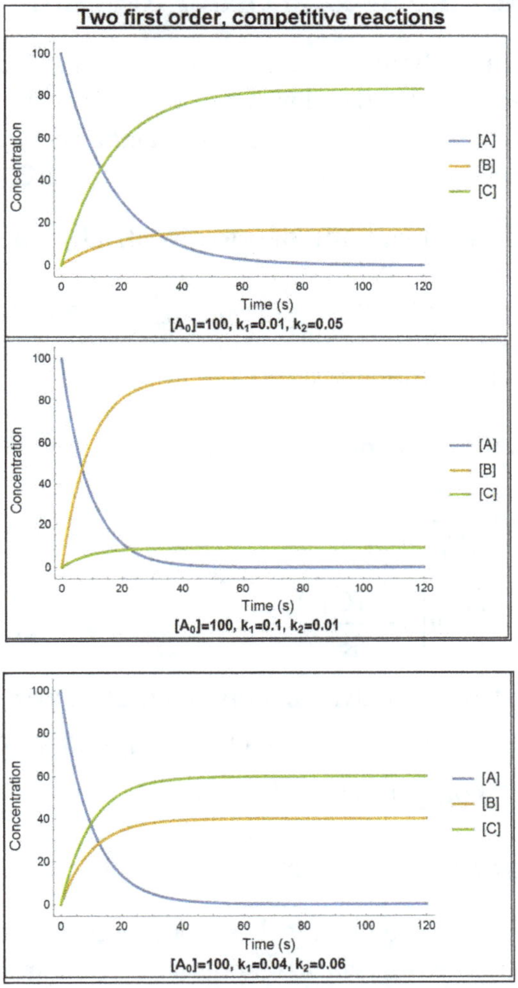

Time course of two first order, competitive reactions with differing rate constants.

When a substance reacts simultaneously to give two different products, a parallel or competitive reaction is said to take place.

Two First Order Reactions

$A \rightarrow B$ and $A \rightarrow C$, with constants k_1 and k_2 and rate equations $-\dfrac{d[A]}{dt} = (k_1 + k_2)[A]$; $\dfrac{d[B]}{dt} = k_1[A]$ and $\dfrac{d[C]}{dt} = k_2[A]$

The integrated rate equations are then $[A] = [A]_0 e^{-(k_1+k_2)t}$; $[B] = \dfrac{k_1}{k_1 + k_2}[A]_0(1 - e^{-(k_1+k_2)t})$ and $[C] = \dfrac{k_2}{k_1 + k_2}[A]_0(1 - e^{-(k_1+k_2)t})$.

One important relationship in this case is $\frac{[B]}{[C]} = \frac{k_1}{k_2}$.

One First Order and One Second Order Reaction

This can be the case when studying a bimolecular reaction and a simultaneous hydrolysis (which can be treated as pseudo order one) takes place: the hydrolysis complicates the study of the reaction kinetics, because some reactant is being "spent" in a parallel reaction. For example, A reacts with R to give our product C, but meanwhile the hydrolysis reaction takes away an amount of A to give B, a byproduct: $\mathrm{A+H_2O \rightarrow B}$ and $A+R \rightarrow C$. The rate equations are: $\frac{d[\mathrm{B}]}{dt} = k_1[\mathrm{A}][\mathrm{H_2O}] = k_1^{'}[\mathrm{A}]$ and $\frac{d[\mathrm{C}]}{dt} = k_2[\mathrm{A}][\mathrm{R}]$. Where $k_1^{'}$ is the pseudo first order constant.

The integrated rate equation for the main product [C] is $[\mathrm{C}] = [\mathrm{R}]_0 \left[1 - e^{-\frac{k_2}{k_1^{'}}[\mathrm{A}]_0(1-e^{-k_1^{'}t})}\right]$, which is equivalent to $\ln\frac{[\mathrm{R}]_0}{[\mathrm{R}]_0 - [\mathrm{C}]} = \frac{k_2[\mathrm{A}]_0}{k_1^{'}}(1-e^{-k_1^{'}t})$. Concentration of B is related to that of C through $[\mathrm{B}] = -\frac{k_1^{'}}{k_2} ln\left(1 - \frac{[\mathrm{C}]}{[\mathrm{R}]_0}\right)$

The integrated equations were analytically obtained but during the process it was assumed that $[\mathrm{A}]_0 - [\mathrm{C}] \approx [\mathrm{A}]_0$ therefeore, previous equation for [C] can only be used for low concentrations of [C] compared to $[\mathrm{A}]_0$.

Stoichiometric Reaction Networks

The most general description of a chemical reaction network considers a number *N* of distinct chemical species reacting via *R* reactions. The chemical equation of the *j*-th reaction can then be written in the generic form:

$$s_{1j}\,\mathrm{X}_1 + s_{2j}\,\mathrm{X}_2 \ldots + s_{Nj}\,\mathrm{X}_N \xrightarrow{k_j} r_{1j}\,\mathrm{X}_1 + r_{2j}\,\mathrm{X}_2 + \ldots + r_{Nj}\,\mathrm{X}_N,$$

which is often written in the equivalent form:

$$\sum_{i=1}^{N} s_{ij}\,\mathrm{X}_i \rightarrow [k_j] \sum_{i=1}^{N} r_{ij}\,\mathrm{X}_i .$$

Here,

- j is the reaction index running from 1 to R,
- X_i denotes the i-th chemical species,

- k_j is the rate constant of the j-th reaction,
- s_{ij} and r_{ij} are the stoichiometric coefficients of reactants and products, respectively.

The rate of such reaction can be inferred by the law of mass action,

$$f_j([\vec{X}]) = k_j \prod_{z=1}^{N} [X_z]^{s_{zj}}$$

which denotes the flux of molecules per unit time and unit volume. Here $[\vec{X}] = ([X_1],[X_2],...,[X_N])$ is the vector of concentrations. This definition includes the elementary reactions:

- Zero order reactions,

for which $s_{zj} = 0$ for all z,

- First order reactions,

for which $s_{zj} = 1$ for a single z,

- Second order reactions,

for which $s_{zj} = 1$ for exactly two z, i.e., a bimolecular reaction, or $s_{zj} = 2$ for a single z, i.e., a dimerization reaction.

One can define the stoichiometric matrix

$$S_{ij} = r_{ij} - s_{ij}$$

denoting the net extent of molecules of *i* in reaction *j*. The reaction rate equations can then be written in the general form

$$\frac{d[X_i]}{dt} = \sum_{j=1}^{R} S_{ij} f_j([\vec{X}]).$$

This is the product of the stoichiometric matrix and the vector of reaction rate functions. Particular simple solutions exist in equilibrium, $\frac{d[X_i]}{dt} = 0$, for systems composed of merely reversible reactions. In this case the rate of the forward and backward reactions are equal, a principle called detailed balance. Detailed balance is a property of the stoichiometric matrix S_{ij} alone and does not depend on the particular form of the rate functions f_j. All other cases where detailed balance is violated are commonly studied by flux balance analysis which has been developed to understand metabolic pathways.

General Dynamics of Unimolecular Conversion

For a general unimolecular reaction involving interconversion of *N* different species, whose concentrations at time *t* are denoted by $X_1(t)$ through $X_N(t)$, an analytic form for the time-evolution of the species can be found. Let the rate constant of conversion from species X_i to species X_j be denoted as k_{ij}, and construct a rate-constant matrix *K* whose entries are the k_{ij}.

Also, let $X(t) = (X_1(t), X_2(t), ..., X_N(t))^T$ be the vector of concentrations as a function of time.

Let $J = (1,1,1,...,1)^T$ be the vector of ones.

Let *I* be the $N \times N$ identity matrix.

Let Diag be the function that takes a vector and constructs a diagonal matrix whose on-diagonal entries are those of the vector.

Let $\mathcal{L}^{-1}$ be the inverse Laplace transform from *s* to *t*.

Then the time-evolved state $X(t)$ is given by:

$$X(t) = \mathcal{L}^{-1}[(sI + Diag(KJ) - K^T)^{-1} X(0)],$$

thus providing the relation between the initial conditions of the system and its state at time *t*.

Arrhenius Equation

In physical chemistry, the Arrhenius equation is a formula for the temperature dependence of reaction rates. The equation was proposed by Svante Arrhenius in 1889, based on the work of Dutch chemist Jacobus Henricus van 't Hoff who had noted in 1884 that van 't Hoff equation for the temperature dependence of equilibrium constants suggests such a formula for the rates of both forward and reverse reactions. This equation has a vast and important application in determining rate of chemical reactions and for calculation of energy of activation. Arrhenius provided a physical justification and interpretation for the formula. Currently, it is best seen as an empirical relationship. It can be used to model the temperature variation of diffusion coefficients, population of crystal vacancies, creep rates, and many other thermally-induced processes/reactions. The Eyring equation, developed in 1935, also expresses the relationship between rate and energy.

Equation:

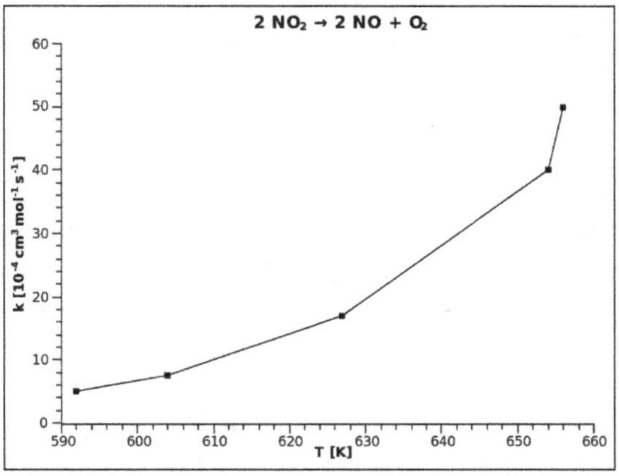

In almost all practical cases, $E_a \gg RT$ and k increases rapidly with T.

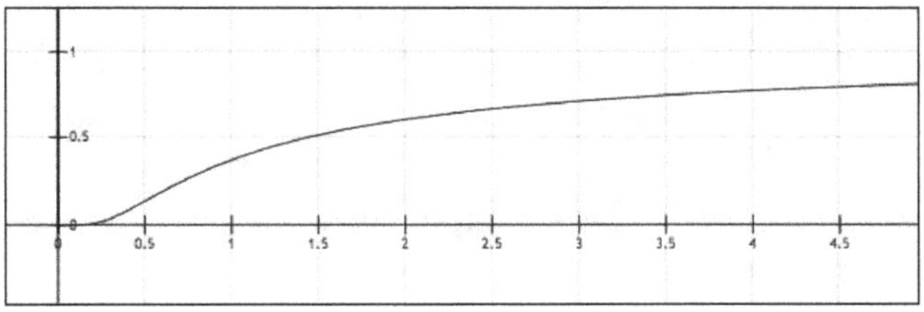

Mathematically, at very high temperatures so that $E_a \gg RT$, k levels off and approaches A as a limit, but this case does not occur under practical conditions.

Arrhenius equation gives the dependence of the rate constant of a chemical reaction on the absolute temperature, a pre-exponential factor and other constants of the reaction.

$$k = Ae^{\frac{-E_a}{RT}},$$

where ,

k is the rate constant,

T is the absolute temperature (in Kelvin),

A is the pre-exponential factor, a constant for each chemical reaction. According to collision theory, A is the frequency of collisions in the correct orientation,

E_a is the activation energy for the reaction (in the same units as RT),

R is the universal gas constant.

Alternatively, the equation may be expressed as:

$$k = Ae^{\frac{-E_a}{k_B T}}$$

where,

E_a is the activation energy for the reaction (in the same units as k_BT),

k_B is the Boltzmann constant.

The only difference is the energy units of E_a: the former form uses energy per mole, which is common in chemistry, while the latter form uses energy per molecule directly, which is common in physics. The different units are accounted for in using either the gas constant, R, or the Boltzmann constant, k_B, as the multiplier of temperature T.

The units of the pre-exponential factor A are identical to those of the rate constant and will vary depending on the order of the reaction. If the reaction is first order it has the units: s^{-1}, and for that reason it is often called the *frequency factor* or *attempt frequency* of the reaction. Most simply, k is the number of collisions that result in a reaction per second, A is the number of collisions (leading to a reaction or not) per second occurring with the proper orientation to react and $e^{-E_a/(RT)}$ is the probability that any given collision will result in a reaction. It can be seen that either increasing the temperature or decreasing the activation energy (for example through the use of catalysts) will result in an increase in rate of reaction.

Given the small temperature range of kinetic studies, it is reasonable to approximate the activation energy as being independent of the temperature. Similarly, under a wide range of practical conditions, the weak temperature dependence of the pre-exponential factor is negligible compared to the temperature dependence of the $\exp(-E_a/(RT))$ factor; except in the case of "barrierless" diffusion-limited reactions, in which case the pre-exponential factor is dominant and is directly observable.

Arrhenius Plot

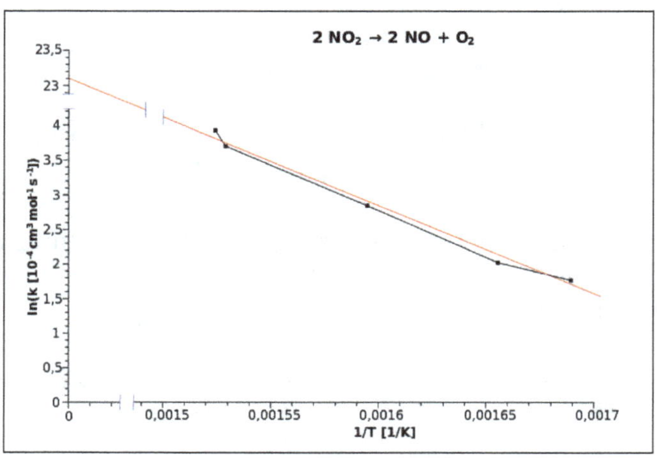

Arrhenius linear plot: ln k against $1/T$.

Taking the natural logarithm of Arrhenius equation yields:

$$\ln k = \ln A - \frac{E_a}{R}\frac{1}{T}.$$

Rearranging yields:

$$\ln k = \frac{-E_a}{R}\left(\frac{1}{T}\right) + \ln A$$

This has the same form as an equation for a straight line:

$$y = mx + b$$

where x is the reciprocal of T.

So, when a reaction has a rate constant that obeys Arrhenius equation, a plot of ln k versus T^{-1} gives a straight line, whose gradient and intercept can be used to determine E_a and A. This procedure has become so common in experimental chemical kinetics that practitioners have taken to using it to *define* the activation energy for a reaction. That is the activation energy is defined to be $(-R)$ times the slope of a plot of ln k vs. $(1/T)$:

$$E_a \equiv -R\left[\frac{\partial \ln k}{\partial(1/T)}\right]_P.$$

Modified Arrhenius Equation

The modified Arrhenius equation makes explicit the temperature dependence of the pre-exponential factor. The modified equation is usually of the form:

$$k = AT^n e^{-E_a/(RT)}.$$

The original Arrhenius expression above corresponds to $n = 0$. Fitted rate constants typically lie in the range $-1 < n < 1$. Theoretical analyses yield various predictions for n. It has been pointed out that "it is not feasible to establish, on the basis of temperature studies of the rate constant, whether the predicted $T^{1/2}$ dependence of the pre-exponential factor is observed experimentally". However, if additional evidence is available, from theory and from experiment (such as density dependence), there is no obstacle to incisive tests of the Arrhenius law.

Another common modification is the stretched exponential form:

$$k = A\exp\left[-\left(\frac{E_a}{RT}\right)^{\beta}\right],$$

where β is a dimensionless number of order 1. This is typically regarded as a purely empirical correction or *fudge factor* to make the model fit the data, but can have theoretical meaning, for example showing the presence of a range of activation energies or in special cases like the Mott variable range hopping.

Theoretical Interpretation of the Equation

Arrhenius's Concept of Activation Energy

Arrhenius argued that for reactants to transform into products, they must first acquire a minimum amount of energy, called the activation energy E_a. At an absolute temperature T, the fraction of molecules that have a kinetic energy greater than E_a can be calculated from statistical mechanics. The concept of *activation energy* explains the exponential nature of the relationship, and in one way or another, it is present in all kinetic theories.

The calculations for reaction rate constants involve an energy averaging over a Maxwell–Boltzmann distribution with E_a as lower bound and so are often of the type of incomplete gamma functions, which turn out to be proportional to $e^{\frac{-E_a}{RT}}$.

Collision Theory

One approach is the collision theory of chemical reactions, developed by Max Trautz and William Lewis in the years 1916–18. In this theory, molecules are supposed to react if they collide with a relative kinetic energy along their line of centers that exceeds E_a. The number of binary collisions between two unlike molecules per second per unit volume is found to be,

$$Z_{AB} = N_A N_B d_{AB}^2 \sqrt{\frac{8\pi k_B T}{\mu_{AB}}},$$

where N_A and N_B are the number of molecules of A and B per unit volume, d_{AB} is the average diameter of A and B, k_B is the Boltzmann constant, and μ is the reduced mass.

The rate constant is then calculated as $k = Z_{AB} e^{\frac{-E_a}{RT}}$ so that the collision theory predicts that the pre-exponential factor is equal to the collision number Z_{AB}. However for many reactions this agrees poorly with experiment, so the rate constant is written instead as $k = PZ_{AB} e^{\frac{-E_a}{RT}}$. Here P is an empirical steric factor, often much less than 1, which is interpreted as the fraction of sufficiently energetic collisions in which the two molecules have the correct mutual orientation to react.

Transition State Theory

The Eyring equation, another Arrhenius-like expression, appears in the "transition state theory" of chemical reactions, formulated by Wigner, Eyring, Polanyi and Evans in the 1930s. The Eyring equation can be written:

$$k = \frac{k_B T}{h} e^{-\frac{\Delta G^\ddagger}{RT}} = \frac{k_B T}{h} e^{\frac{\Delta S^\ddagger}{R}} e^{-\frac{\Delta H^\ddagger}{RT}},$$

where $\Delta G^{\ddagger}$ = the Gibbs energy of activation, $\Delta S^{\ddagger}$ is the entropy of activation, $\Delta H^{\ddagger}$ is the enthalpy of activation, k_B is Boltzmann's constant, and h is Planck's constant.

At first sight this looks like an exponential multiplied by a factor that is *linear* in temperature. However, free energy is itself a temperature dependent quantity. The free energy of activation $\Delta G^{\ddagger} = \Delta H^{\ddagger} - T\Delta S^{\ddagger}$ is the difference of an enthalpy term and an entropy term multiplied by the absolute temperature. The pre-exponential factor depends primarily on the entropy of activation. The overall expression again takes the form of an Arrhenius exponential (of enthalpy rather than energy) multiplied by a slowly varying function of T. The precise form of the temperature dependence depends upon the reaction, and can be calculated using formulas from statistical mechanics involving the partition functions of the reactants and of the activated complex.

Limitations of the Idea of Arrhenius Activation Energy

Both the Arrhenius activation energy and the rate constant k are experimentally determined, and represent macroscopic reaction-specific parameters that are not simply related to threshold energies and the success of individual collisions at the molecular level. Consider a particular collision (an elementary reaction) between molecules A and B. The collision angle, the relative translational energy, the internal (particularly vibrational) energy will all determine the chance that the collision will produce a product molecule AB. Macroscopic measurements of E and k are the result of many individual collisions with differing collision parameters. To probe reaction rates at molecular level, experiments are conducted under near-collisional conditions and this subject is often called molecular reaction dynamics.

There are deviations from the Arrhenius law during the glass transition in all classes of glass-forming matter. The Arrhenius law predicts that the motion of the structural units (atoms, molecules, ions, etc.) should slow down at a slower rate through the glass transition than is experimentally observed. In other words, the structural units slow down at a faster rate than is predicted by the Arrhenius law. This observation is made reasonable assuming that the units must overcome an energy barrier by means of a thermal activation energy. The thermal energy must be high enough to allow for translational motion of the units which leads to viscous flow of the material.

Collision Theory

Collision theory is a theory that was proposed independently by Max Trautz in 1916 and William Lewis in 1918. The collision theory states that when suitable particles of the reactant hit each other, only a certain fraction of the collisions cause any noticeable or significant chemical change; these successful changes are called successful collisions. The successful collisions must have enough energy, also known as activation energy, at the moment of impact to break the pre-existing bonds and form all new bonds. This results in the products of the reaction. Increasing the concentration of the reactant particles or raising the temperature - which brings about more collisions and hence more successful collisions - therefore increases the rate of a reaction.

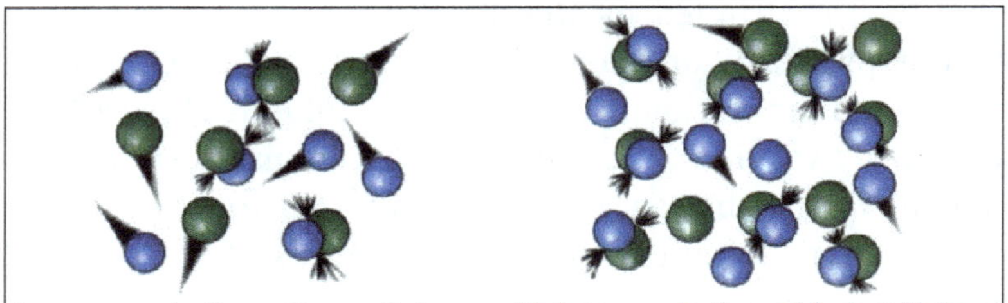

Reaction rate tends to increase with concentration phenomenon explained by collision theory.

When a catalyst is involved in the collision between the reactant molecules, less energy is required for the chemical change to take place, and hence more collisions have sufficient energy for reaction to occur. The reaction rate therefore increases.

Collision theory is closely related to chemical kinetics.

Rate Constant

The rate for a bimolecular gas-phase reaction, predicted by collision theory is:

$$r(T) = kn_A n_B = Z\rho \exp\left(\frac{-E_a}{RT}\right)$$

unit is number of molecular reactions s^{-1} m^{-3}.

where:

- k is the rate constant in unit (number of molecules)$^{-1}$ $s^{-1}m^3$.
- n_A is the number density of A in the gas (unit m^{-3}).
- n_B is the number density of B in the gas (unit m^{-3}). E.g. for a gas mixture with gas A concentration 0.1 mol L^{-1} and B concentration 0.2 mol L^{-1}, the number of density of A is $0.1\times6.02\times10^{23}\div10^{-3} = 6.02\times10^{25}$ m^{-3}, the number of density of B is $0.2\times6.02\times10^{23}\div10^{-3} = 1.2\times10^{26}$ m^{-3}.
- Z is the collision frequency (unit $m^{-3}s^{-1}$).
- ρ is the steric factor.
- E_a is the activation energy of the reaction, (unit J mol^{-1}).

- T is the temperature (unit K).
- R is the gas constant (unit J mol^{-1}K^{-1}).

The unit of $r(T)$ can be converted to mol L^{-1}s^{-1}, after divided by ($1000\times N_A$), where N_A is the Avogadro constant.

For a reaction between A and B, the collision frequency is:

$$Z = n_A n_B \sigma_{AB}\sqrt{\frac{8k_BT}{\pi\mu_{AB}}} = 10^6 N_A^2[A][B]\sigma_{AB}\sqrt{\frac{8k_BT}{\pi\mu_{AB}}}$$

The SI unit of Z is number of collision m^{-3}s^{-1}.

where:

- σ_{AB} is the reaction cross section (unit m^2), the area when two molecules collide with each other, simplified to $\sigma_{AB} = \pi(r_A + r_B)^2$, where r_A the radius of A and r_B the radius of B in unit m. This is not the true reaction picture but rather is over simplified. The effective area should be smaller than σ_{AB}. This difference can be corrected by the reaction steric factor in the rate equation.
- k_B is the Boltzmann's constant unit J K^{-1}.
- T is the absolute temperature (unit K).
- μ_{AB} is the reduced mass of the reactants A and B, $\mu_{AB} = \frac{m_A m_B}{m_A + m_B}$ (unit kg).
- [A] is molar concentration of A in unit mol L^{-1}.
- [B] is molar concentration of B in unit mol L^{-1}.

If all the units that are related to dimension are converted to dm, i.e. mol dm^{-3} for [A] and [B], dm^2 for σ_{AB}, dm^{2}kg s^{-2}K^{-1} for Boltzmann's constant, then

$$Z = N_A \sigma_{AB}\sqrt{\frac{8k_BT}{\pi\mu_{AB}}}[A][B]$$

unit mol dm^{-3} s^{-1}.

Collision in Diluted Solution

Collision in diluted gas or liquid solution is regulated by diffusion instead of direct collisions, which can be calculated from Fick's laws of diffusion.

For a diluted solution in the gas or the liquid phase, the above equation is not suitable when diffusion takes control of the collision frequency, i.e., The direct collision between the two molecules no longer dominates. For any given molecule A, it has to collide with a lot of solvent molecules, let's say molecule C, before finding the B molecule to react with.

Quantitative Insights

Derivation

Consider the bimolecular elementary reaction:

$$A + B \rightarrow C$$

In collision theory it is considered that two particles A and B will collide if their nuclei get closer than a certain distance. The area around a molecule A in which it can collide with an approaching B molecule is called the cross section (σ_{AB}) of the reaction and is, in simplified terms, the area corresponding to a circle whose radius (r_{AB}) is the sum of the radii of both reacting molecules, which are supposed to be spherical. A moving molecule will therefore sweep a volume $\pi r_{AB}^2 c_A$ per second as it moves, where c_A is the average velocity of the particle. (This solely represents the classical notion of a collision of solid balls. As molecules are quantum-mechanical many-particle systems of electrons and nuclei based upon the Coulomb and exchange interactions, generally they neither obey rotational symmetry nor do they have a box potential. Therefore, more generally the cross section is defined as the reaction probability of a ray of A particles per areal density of B targets, which makes the definition independent from the nature of the interaction between A and B. Consequently, the radius r_{AB} is related to the length scale of their interaction potential).

From kinetic theory it is known that a molecule of A has an average velocity (different from root mean square velocity) of $c_A = \sqrt{\frac{8k_B T}{\pi m_A}}$, where k_B is Boltzmann constant, and m_A is the mass of the molecule.

The solution of the two-body problem states that two different moving bodies can be treated as one body which has the reduced mass of both and moves with the velocity of the center of mass, so, in this system μ_{AB} must be used instead of m_A. Thus, for a given molecule A, it travels $t = l / c_A = 1/(n_B \sigma_{AB} c_A)$ before hitting a molecule B if all B is fixed with no movement, where l is the average traveling distance. Since B also moves, the relative velocity can be calculated using the reduced mass of A and B.

Therefore, the total collision frequency, of all A molecules, with all B molecules, is:

$$Z = n_A n_B \sigma_{AB} \sqrt{\frac{8k_B T}{\pi \mu_{AB}}} = 10^6 N_A^2 [A][B] \sigma_{AB} \sqrt{\frac{8k_B T}{\pi \mu_{AB}}} = z[A][B],$$

From Maxwell–Boltzmann distribution it can be deduced that the fraction of collisions with more energy than the activation energy is $e^{\frac{-E_a}{RT}}$. Therefore, the rate of a bimolecular reaction for ideal gases will be,

$$r = z\rho[A][B]\exp\left(\frac{-E_a}{RT}\right),$$

In unit number of molecular reactions $s^{-1}m^{-3}$.

Where:

- Z is the collision frequency with unit $s^{-1}m^{-3}$. The z is Z without [A][B].
- ρ is the steric factor.
- E_a is the activation energy (per mole) of the reaction in unit J/mol.
- T is the absolute temperature in unit K.
- R is the gas constant in unit J/mol/K.
- [A] is molar concentration of A in unit mol/L.
- [B] is molar concentration of B in unit mol/L.

The product $z\rho$ is equivalent to the preexponential factor of the Arrhenius equation.

Validity of the Theory and Steric Factor

Once a theory is formulated, its validity must be tested, that is, compare its predictions with the results of the experiments.

When the expression form of the rate constant is compared with the rate equation for an elementary bimolecular reaction, $r = k(T)[A][B]$, it is noticed that

$$k(T) = N_A\sigma_{AB}\rho\sqrt{\frac{8k_BT}{\pi\mu_{AB}}}\exp\left(\frac{-E_a}{RT}\right)$$

unit $M^{-1}s^{-1}$ (= $dm^3\ mol^{-1}s^{-1}$), with all dimension unit dm including k_B.

This expression is similar to the Arrhenius equation and gives the first theoretical explanation for the Arrhenius equation on a molecular basis. The weak temperature dependence of the preexponential factor is so small compared to the exponential factor that it cannot be measured experimentally, that is, "it is not feasible to establish, on the basis of temperature studies of the rate constant, whether the predicted $T^{1/2}$ dependence of the preexponential factor is observed experimentally".

Steric Factor

If the values of the predicted rate constants are compared with the values of known rate constants, it is noticed that collision theory fails to estimate the constants correctly, and the more complex the molecules are, the more it fails. The reason for this is that particles have been supposed to be spherical and able to react in all directions, which is not true, as the orientation of the collisions is not always proper for the reaction. For example, in the hydrogenation reaction of ethylene the H_2 molecule must approach the bonding zone between the atoms, and only a few of all the possible collisions fulfill this requirement.

To alleviate this problem, a new concept must be introduced: the steric factor ρ. It is defined as the ratio between the experimental value and the predicted one (or the ratio between the frequency factor and the collision frequency):

$$\rho = \frac{A_{\text{observed}}}{Z_{\text{calculated}}},$$

and it is most often less than unity.

Usually, the more complex the reactant molecules, the lower the steric factor. Nevertheless, some reactions exhibit steric factors greater than unity: the harpoon reactions, which involve atoms that exchange electrons, producing ions. The deviation from unity can have different causes: the molecules are not spherical, so different geometries are possible; not all the kinetic energy is delivered into the right spot; the presence of a solvent (when applied to solutions), etc.

Experimental rate constants compared to the ones predicted by collision theory for gas phase reactions			
Reaction	A, $s^{-1}M^{-1}$	Z, $s^{-1}M^{-1}$	Steric factor
$2\,ClNO \rightarrow 2\,Cl + 2\,NO$	9.4×10^{9}	5.9×10^{10}	0.16
$2\,ClO \rightarrow Cl_2 + O_2$	6.3×10^{7}	2.5×10^{10}	2.3×10^{-3}
$H_2 + C_2H_4 \rightarrow C_2H_6$	1.24×10^{6}	7.3×10^{11}	1.7×10^{-6}
$Br_2 + K \rightarrow KBr + Br$	1.0×10^{12}	2.1×10^{11}	4.3

Collision theory can be applied to reactions in solution; in that case, the *solvent cage* has an effect on the reactant molecules, and several collisions can take place in a single encounter, which leads to predicted preexponential factors being too large. ρ values greater than unity can be attributed to favorable entropic contributions.

Experimental rate constants compared to the ones predicted by collision theory for reactions in solution				
Reaction	Solvent	A, 10^{11} $s^{-1}M^{-1}$	Z, 10^{11} $s^{-1}M^{-1}$	Steric factor
$C_2H_5Br + OH^-$	ethanol	4.30	3.86	1.11
$C_2H_5O^- + CH_3I$	ethanol	2.42	1.93	1.25
$ClCH_2CO_2^- + OH^-$	water	4.55	2.86	1.59
$C_3H_6Br_2 + I^-$	methanol	1.07	1.39	0.77
$HOCH_2CH_2Cl + OH^-$	water	25.5	2.78	9.17
$4\text{-}CH_3C_6H_4O^- + CH_3I$	ethanol	8.49	1.99	4.27
$CH_3(CH_2)_2Cl + I^-$	acetone	0.085	1.57	0.054
$C_5H_5N + CH_3I$	$C_2H_2Cl_4$	—	—	2.0 10×10^{-6}

Catalysis

In chemistry, Catalysis is the modification of the rate of a chemical reaction, usually an acceleration, by addition of a substance not consumed during the reaction. The rates of chemical reactions—that is, the velocities at which they occur—depend upon a number of factors, including the chemical nature of the reacting species and the external conditions to which they are exposed. A particular phenomenon associated with the rates of chemical reactions that is of great theoretical and practical interest is catalysis, the acceleration of chemical reactions by substances not consumed in the reactions themselves—substances known as catalysts. The study of catalysis is of interest theoretically because of what it reveals about the fundamental nature of chemical reactions; in practice, the study of catalysis is important because many industrial processes depend upon catalysts for their success. Fundamentally, the peculiar phenomenon of life would hardly be possible without the biological catalysts termed enzymes.

In a catalyzed reaction, the catalyst generally enters into chemical combination with the reactants but is ultimately regenerated, so the amount of catalyst remains unchanged. Since the catalyst is not consumed, each catalyst molecule may induce the transformation of many molecules of reactants. For an active catalyst, the number of molecules transformed per minute by one molecule of catalyst may be as large as several million.

Where a given substance or a combination of substances undergoes two or more simultaneous reactions that yield different products, the distribution of products may be influenced by the use of a catalyst that selectively accelerates one reaction relative to the others. By choosing the appropriate catalyst, a particular reaction can be made to occur to the extent of practically excluding another. Many important applications of catalysis are based on selectivity of this kind.

Since a reverse chemical reaction may proceed by reversal of the steps constituting the mechanism of the forward reaction, the catalyst for a given reaction accelerates the reaction in both directions equally. Therefore, a catalyst does not affect the position of equilibrium of a chemical reaction; it affects only the rate at which equilibrium is attained. Apparent exceptions to this generalization are those reactions in which one of the products is also a catalyst for the reaction. Such reactions are termed autocatalytic.

Cases are also known in which the addition of a foreign substance, called an inhibitor, decreases the rate of a chemical reaction. This phenomenon, properly termed inhibition or retardation, is sometimes called negative catalysis. Concentrations of the inhibitor may in some cases be much lower than those of the reactants. Inhibition may result from (1) a decrease in the concentration of one of the reactants because of complex formation between the reactant and the inhibitor, (2) a decrease in the concentration of an active catalyst ("poisoning" of the catalyst) because of complex formation between the catalyst and the inhibitor, or (3) a termination of a chain reaction because of destruction of the chain carriers by the inhibitor.

Catalysts

For a chemical reaction to take place, it requires a certain minimum amount of energy, called its activation energy. If a substance can lower this activation energy without itself being changed or consumed during the reaction, it is called a catalyst or catalytic agent. The action of a catalyst is called catalysis.

Catalytic converter on a Saab 9-5.

The catalyst reduces the activation energy by providing an alternative pathway for the reaction to occur. In so doing, the catalytic agent makes the reaction proceed faster and at a lower temperature than the uncatalyzed reaction. In addition, if a given reaction can produce a variety of products, the catalyst may help direct the reaction toward generating a specific subgroup of products.

Catalysts play essential roles in biological systems and industrial processes. In living organisms, enzymes catalyze numerous biochemical reactions that sustain life and are involved in growth and reproductive functions. In industry, catalysts are used for a wide array of manufacturing processes, including the production of ammonia, sulfuric acid, polymers, and petroleum product substitutes. In industrialized countries, perhaps the most familiar example of the use of catalysts lies in the catalytic converter, a device that breaks down harmful substances in automobile exhaust. Research into catalysis and the use of catalysts continues to be a focus of many investigators in the pure and applied sciences.

In a more general sense, the term catalyst may be applied to any agent (including a person or group) that brings about accelerated change. For example, someone may be called a "catalyst for political change."

General Catalytic Process

A catalyst participates in one or more stages of a reaction, but it is usually not a reactant or product of the overall reaction that it catalyzes. An exception to this rule is the process known as *autocatalysis*, in which the reaction product acts as a catalyst for the reaction. A substance that inhibits the action of a catalyst is called an *inhibitor*; one that accelerates the action of a catalyst is called a *promoter*.

A catalyst may react with one or more reactants to form a chemical intermediate, and this intermediate subsequently reacts to form the final reaction product. In the overall process, the catalyst is regenerated. Alternatively, the catalyst may provide a surface to which the reactants bind, facilitating their reaction by bringing them close together. The products that are formed are released from the catalyst.

Consider the following reaction scheme, in which C represents the catalyst, A and B are reactants, and D is the product of the reaction of A and B.

$$\mathrm{A+C} \rightarrow \mathrm{AC}\ (1)$$

$$\mathrm{B+AC} \rightarrow \mathrm{ABC}\ (2)$$

$$\mathrm{ABC} \rightarrow \mathrm{CD}\ (3)$$

$$\mathrm{CD} \rightarrow \mathrm{C+D}\ (4)$$

Here, the catalyst (C) is consumed by the reaction in stage one, but it is regenerated in stage four. Thus, the overall reaction can be written as:

$$\mathrm{A+B+C} \rightarrow \mathrm{D+C}$$

Types of Catalysts

Catalysts can be either heterogeneous or homogeneous. Biological catalysts (or biocatalysts) are often considered a separate group.

A heterogeneous catalyst is one that is in a different phase from that of the reactants. For example, a solid catalyst may be used in a liquid reaction mixture. On the other hand, a homogeneous catalyst is one that is in the same phase as that of the reactants. For example, the catalyst may be dissolved in a liquid reaction mixture.

Heterogeneous Catalysts

A simple model for heterogeneous catalysis involves the catalyst providing a surface on which the reactants (or substrates) temporarily become adsorbed. Chemical bonds in the substrate become sufficiently weakened for new bonds to be created. As the products are generated, they bind relatively weakly to the catalyst and are therefore released. Different possible mechanisms for reactions on surfaces are known, depending on how the adsorption takes place.

For example, consider the Haber process to manufacture ammonia from nitrogen and hydrogen. In this case, finely divided iron acts as a heterogeneous catalyst. As the reactant molecules (hydrogen and nitrogen) bind to the catalyst, this binding process has two effects: first, the molecules come closer together than they would be in the gas phase; and second, their internal bonds are weakened. In this manner, the catalyst makes it possible for the reactant molecules to react faster than they would if they had remained in the gas phase.

Homogeneous Catalysts

In homogeneous catalysis, the catalyst itself may be transformed at an early stage of the reaction, and it is regenerated by the end of the reaction. An example is the breakdown of ozone by chlorine free radicals (free atoms of chlorine). Chlorine free radicals are formed by the action of ultraviolet radiation on chlorofluorocarbons (CFCs). These free radicals react with ozone to form oxygen molecules, and chlorine free radicals are regenerated. Some of the simplest reactions are as follows.

$$\mathrm{Cl\bullet} + \mathrm{O_3} \rightarrow \mathrm{ClO\bullet} + \mathrm{O_2}$$

$$\mathrm{ClO\bullet} + \mathrm{O_3} \rightarrow \mathrm{Cl\bullet} + 2\,\mathrm{O_2}$$

Biological Catalysts

In nature, enzymes are catalysts for biochemical reactions that take place within living organisms. Most enzymes are proteins, but some enzymes—called *ribozymes*—are made of RNA. Some DNA molecules, called *deoxyribozymes*, have been found to have catalytic activity. In addition, some antibodies, usually prepared artificially, have catalytic activity and are called *abzymes*.

Reaction Energetics

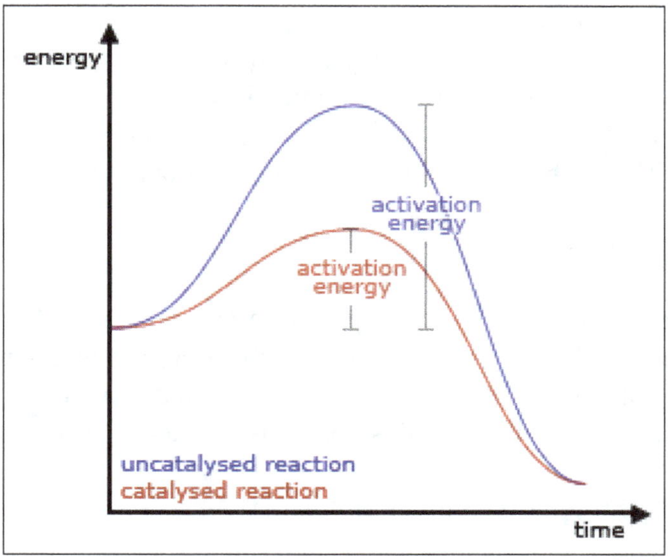

This generic graph shows the effect of a catalyst in a hypothetical chemical reaction. The energy level of the initial reactants is on the extreme left of the graph; that of the final products is on the extreme right. Notice that the catalyzed pathway (red) has a lower activation energy, but it produces the same final result as the uncatalyzed pathway (blue).

Catalysts work by providing an alternative mechanism for a chemical reaction—a mechanism that lowers the activation energy relative to that of the uncatalyzed reaction. This means that catalysts reduce the amount of energy needed to start a chemical reaction. Also, the "transition state" (unstable transitional intermediate) formed by the reactants during a catalyzed reaction is usually different from that formed during an uncatalyzed reaction.

The energy profile diagram on the right shows that the catalyzed pathway (red) has a lower activation energy than the uncatalyzed pathway (blue). In addition, it shows that the net change of energy for the overall reaction is the same, whether a catalyst is used or not.

Thus, catalysts can enable reactions to take place at much faster rates, or in certain specific ways, or at lower temperatures. In addition, some reactions run only in the presence of a catalyst.

Catalysts cannot make energetically unfavorable reactions possible. They have *no* effect on the chemical equilibrium of a reaction, because the rates of the forward and the reverse reactions are equally affected by the catalyst.

The SI derived unit for measuring the "catalytic activity" of a catalyst is the katal (which is moles per second). In biochemistry, the catalytic activity of an enzyme is measured in terms of enzyme units.

The activity of a catalyst can also be described by the *turnover number* (TON). In enzymology, turnover number is defined as the maximum number of moles of substrate that an enzyme can convert to product per catalytic site (of the enzyme) per unit time. In more chemical fields, turnover number (TON) is defined as the number of moles of substrate that a mole of catalyst can convert before the catalyst becomes inactivated. The term turnover frequency (TOF) is used to refer to the turnover per unit time, as in enzymology.

Poisoning of a Catalyst

A catalyst can be poisoned if another compound (similar to an inhibitor) alters it chemically or bonds to it and does not release it. Such interactions effectively destroy the usefulness of the catalyst, as it can no longer participate in the reaction that it was supposed to catalyze. Common catalyst poisons are lead, sulfur, zinc, manganese, and phosphorus.

Applications

According to some estimates, 60 percent of all commercially produced chemical products require catalysts at some stage during their manufacture. The most effective catalysts are usually transition metals or transition metal complexes.

The catalytic converter of an automobile is a well-known example of the use of catalysts. In this device, platinum, palladium, or rhodium may be used as catalysts, as they help break down some of the more harmful byproducts of automobile exhaust. A "three-way" catalytic converter performs three tasks: (a) reduction of nitrogen oxides to nitrogen and oxygen; (b) oxidation of carbon monoxide to carbon dioxide; and (c) oxidation of unburnt hydrocarbons to carbon dioxide and water.

Other examples of catalysts and their applications are as follows.

- Ordinary iron is used as a catalyst in the Haber process to synthesize ammonia from nitrogen and hydrogen.
- The mass production of a polymer such as polyethylene or polypropylene is catalyzed by an agent known as the Ziegler-Natta catalyst, which is based on titanium chloride and alkyl aluminum compounds.
- Vanadium(V) oxide is a catalyst for the manufacture of sulfuric acid at high concentrations, by a method known as the *contact process*.
- Nickel is used in the manufacture of margarine.

- Alumina and silica are catalysts in the breakdown of large hydrocarbon molecules into simpler ones—a process known as *cracking*.
- A number of enzymes are used for chemical transformations of organic compounds. These enzymes are called *biocatalysts* and their action is called *biocatalysis*.
- The electrodes of a fuel cell are coated with a catalyst such as platinum, palladium, or nanoscale iron powder.
- The Fischer-Tropsch process is a chemical reaction in which carbon monoxide and hydrogen are converted into liquid hydrocarbons, in the presence of catalysts based on iron and cobalt. This process is mainly used to produce a synthetic petroleum substitute for fuel or lubrication oil.
- Hydrogenation reactions, which involve the addition of hydrogen to organic compounds such as alkenes or aldehydes, require a catalyst such as platinum, palladium, rhodium, or ruthenium.
- A number of chemical reactions are catalyzed by acids or bases.

Heterogeneous Catalysis

Heterogeneous catalysis is the type of catalysis where the phase of the catalyst differs from the phase of the reactants or products. Contrasts with homogeneous catalysis where the reactants, products and catalyst exist in the same phase. Phase distinguishes between not only solid, liquid, and gas components, but also immiscible mixtures (e.g. oil and water), or anywhere an interface is present. Catalysts are useful because they increase the rate of a reaction without themselves being consumed and are therefore reusable.

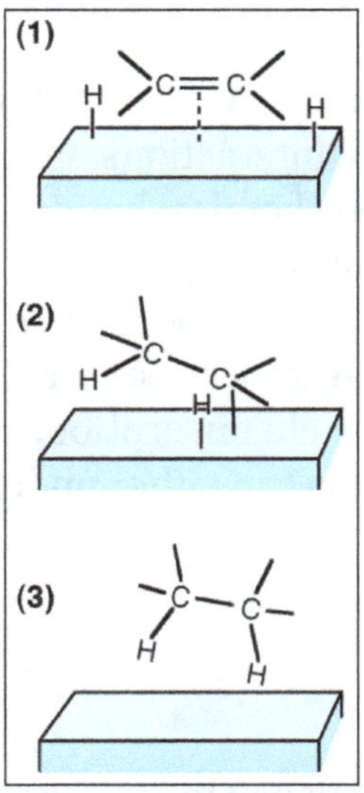

Hydrogenation of ethene on a catalytic solid surface (1) Adsorption (2) Reaction (3) Desorption.

Heterogeneous catalysis typically involves solid phase catalysts and gas phase reactants. In this case, there is a cycle of molecular adsorption, reaction, and desorption occurring at the catalyst surface. Thermodynamics, mass transfer, and heat transfer influence the rate (kinetics) of reaction.

Heterogeneous catalysis is very important because it enables faster, large-scale production and the selective product formation. Approximately 35% of the world's GDP is influenced by catalysis. The production of 90% of chemicals (by volume) is assisted by solid catalysts. The chemical and energy industries rely heavily on heterogeneous catalysis. For example, the Haber-Bosch process uses metal-based catalysts in the synthesis of ammonia, an important component in fertilizer; 144 million tons of ammonia were produced in 2016.

Adsorption

Adsorption is an essential step in heterogeneous catalysis. Adsorption is the process by which a gas (or solution) phase molecule (the adsorbate) binds to solid (or liquid) surface atoms (the adsorbent). The reverse of adsorption is desorption, the adsorbate splitting from adsorbent. In a reaction facilitated by heterogeneous catalysis, the catalyst is the adsorbent and the reactants are the adsorbate.

Types of Adsorption

Two types of adsorption are recognized: physisorption, weakly bound adsorption, and chemisorption, strongly bound adsorption. Many processes in heterogeneous catalysis lie between the two extremes. The Lennard-Jones model provides a basic framework for predicting molecular interactions as a function of atomic separation.

Physisorption

In physisorption, a molecule becomes attracted to the surface atoms via van der Waals forces. These include dipole-dipole interactions, induced dipole interactions, and London dispersion forces. Note that no chemical bonds are formed between adsorbate and adsorbent, and their electronic states remain relatively unperturbed. Typical energies for physisorption are from 3 to 10 kcal/mol. In heterogeneous catalysis, when a reactant molecule physisorbs to a catalyst, it is commonly said to be in a precursor state, an intermediate energy state before chemisorption, a more strongly bound adsorption. From the precursor state, a molecule can either undergo chemisorption, desorption, or migration across the surface. The nature of the precursor state can influence the reaction kinetics.

Chemisorption

When a molecule approaches close enough to surface atoms such that their electron clouds overlap, chemisorption can occur. In chemisorption, the adsorbate and adsorbent share electrons signifying the formation of chemical bonds. Typical energies for chemisorption range from 20 to 100 kcal/mol. Two cases of chemisorption are:

- Molecular adsorption: the adsorbate remains intact. An example is alkene binding by platinum.
- Dissociation adsorption: one or more bonds break concomitantly with adsorption. In this case, the barrier to dissociation affects the rate of adsorption. An example of this the binding of H_2 to a metal catalyst, where the H-H bond is broken upon adsorption.

Surface Reactions

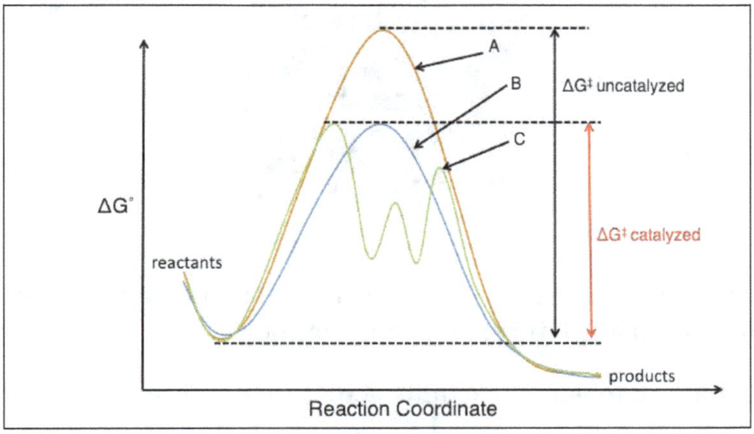

Reaction Coordinate. (A) Uncatalyzed (B) Catalyzed (C) Catalyzed with discreet intermediates (transition states).

Most metal surface reactions occur by chain propagation in which catalytic intermediates are cyclically produced and consumed. Two main mechanisms for surface reactions can be described for $A + B \rightarrow C$.

- Langmuir-Hinshelwood mechanism: The reactant molecules, A and B, both adsorb to the catalytic surface. While adsorbed to the surface, they combine to form product C, which then desorbs.
- Eley-Rideal mechanism: One reactant molecule, A, adsorbs to the catalytic surface. Without adsorbing, B reacts with absorbed A to form C, that then desorbs from the surface.

Most heterogeneously catalyzed reactions are described by the Langmuir-Hinshelwood model.

In heterogeneous catalysis, reactants diffuse from the bulk fluid phase to adsorb to the catalyst surface. The adsorption site is not always an active catalyst site, so reactant molecules must migrate across the surface to an active site. At the active site, reactant molecules will react to form product molecule(s) by following a more energetically facile path through catalytic intermediates. The product molecules then desorb from the surface and diffuse away. The catalyst itself remains intact and free to mediate further reactions. Transport phenomena such as heat and mass transfer, also play a role in the observed reaction rate.

Catalyst Design

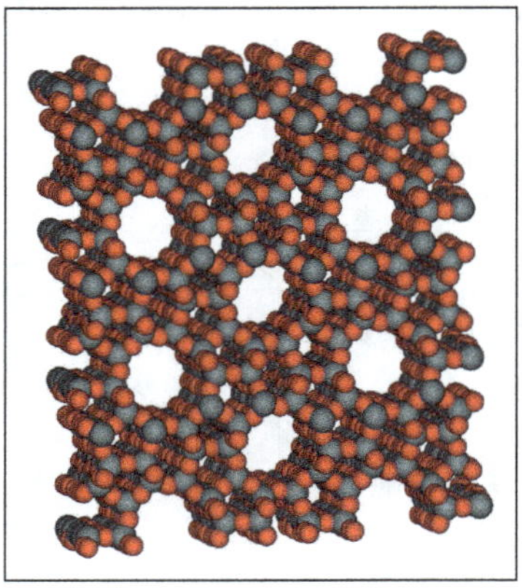

Zeolite structure. A common catalyst support material in hydrocracking. Also acts as a catalyst in hydrocarbon alkylation and isomerization.

Catalysts are not active towards reactants across their entire surface; only specific locations possess catalytic activity, called active sites. The surface area of a solid catalyst has a strong influence on the number of available active sites. In industrial practice, solid catalysts are often porous to maximize surface area, commonly achieving 50–400 m^2/g. Some mesoporous silicates, such as the MCM-41, have surface areas greater than 1000 m^2/g. Porous materials are cost effective due to their high surface area-to-mass ratio and enhanced catalytic activity.

In many cases, a solid catalyst is dispersed on a supporting material to increase surface area (spread the number of active sites) and provide stability. Usually catalyst supports are inert, high melting point materials, but they can also be catalytic themselves. Most catalyst supports are porous (frequently carbon, silica, zeolite, or alumina-based) and chosen for their high surface area-to-mass ratio. For a given reaction, porous supports must be selected such that reactants and products can enter and exit the material.

Often, substances are intentionally added to the reaction feed or on the catalyst to influence catalytic activity, selectivity, and stability. These compounds are called promoters. For example, alumina (Al_2O_3) is added during ammonia synthesis to providing greater stability by slowing sintering processes on the Fe-catalyst.

Sabatier principle can be considered as one of the cornerstones of modern theory of catalysis. Sabatier principle states that the surface-adsorbates interaction has to be an optimal amount: not too weak to be inert toward the reactants and not too strong to poison the surface and avoid desorption of the products. The statement that the surface-adsorbate interaction has to be an optimum, is a qualitative one. Usually the number of adsorbates and transition states associated with a chemical reaction is a large number, thus the optimum has to be found in a many-dimensional space. Catalyst design in such a many-dimensional space is not a computationally viable task. Additionally, such optimization process would be far from intuitive. Scaling relations are used to decrease the dimensionality of the space of catalyst design. Such relations are correlations among adsorbates binding energies (or among adsorbate binding energies and transition states also known as BEP relations) that are "similar enough" e.g., OH versus OOH scaling.

Applying scaling relations to the catalyst design problems greatly reduces the space dimensionality (sometimes to as small as 1 or 2). One can also use micro-kinetic modeling based on such scaling relations to take into account the kinetics associated with adsorption, reaction and desorption of molecules under specific pressure or temperature conditions. Such modeling then leads to well-known volcano-plots at which the optimum qualitatively described by the Sabatier principle is referred to as the "top of the volcano". Scaling relations can be used not only to connect the energetics of radical surface-adsorbed groups (e.g., O*,OH*), but also to connect the energetics of closed-shell molecules among each other or to the counterpart radical adsorbates. A recent challenge for researchers in catalytic sciences is to "break" the scaling relations. The correlations which are manifested in the scaling relations confine the catalyst design space, preventing one from reaching the "top of the volcano".

Breaking scaling relations can refer to either designing surfaces or motifs that do not follow a scaling relation, or ones that follow a different scaling relation (than the usual relation for the associated adsorbates) in the right direction: one that can get us closer to the top of the reactivity volcano. In addition to studying catalytic reactivity, scaling relations can be used to study and screen materials for selectivity toward a special product. There are special combination of binding energies that favor specific products over the others. Sometimes a set of binding energies that can change the selectivity toward a specific product "scale" with each other, thus to improve the selectivity one has to break some scaling relations; an example of this is the scaling between methane and methanol oxidative activation energies that leads to the lack of selectivity in direct conversion of methane to methanol.

Catalyst Deactivation

Catalyst deactivation is defined as a loss in catalytic activity and selectivity over time.

Substances that decrease reaction rate are called poisons. Poisons chemisorb to catalyst surface and reduce the number of available active sites for reactant molecules to bind to. Common poisons include Group V, VI, and VII elements (e.g. S, O, P, Cl), some toxic metals (e.g. As, Pb), and adsorbing species with multiple bonds (e.g. CO, unsaturated hydrocarbons). For example, sulfur disrupts the production of methanol by poisoning the Cu/ZnO catalyst. Substances that increase reaction rate are called promoters. For example, the presence of alkali metals in ammonia synthesis increases the rate of N_2 dissociation.

The presence of poisons and promoters can alter the activation energy of the rate-limiting step and affect a catalyst's selectivity for the formation of certain products. Depending on the amount, a substance can be favorable or unfavorable for a chemical process. For example, in the production of ethylene, a small amount of chemisorbed chlorine will act as a promoter by improving Ag-catalyst selectivity towards ethylene over CO_2, while too much chlorine will act as a poison.

Other mechanisms for catalyst deactivation include:

- Sintering: When heated, dispersed catalytic metal particles can migrate across the support surface and form crystals. This results in a reduction of catalyst surface area.
- Fouling: The deposition of materials from the fluid phase onto the solid phase catalyst and support surfaces. This results in active site and pore blockage.
- Coking: The deposition of heavy, carbon-rich solids onto surfaces due to the decomposition of hydrocarbons.
- Vapor-solid reactions: Formation of an inactive surface layer and formation of a volatile compound that exits the reactor. This results in a loss of surface area and catalyst material.
- Solid-state transformation: Solid-state diffusion of catalyst support atoms to the surface followed by a reaction that forms an inactive phase. This results in a loss of catalyst surface area.
- Erosion: Continual attrition of catalyst material common in fluidized-bed reactors. This results in a loss of catalyst material.

In industry, catalyst deactivation costs billions every year due to process shutdown and catalyst replacement.

Industrial Examples

In industry, many design variables must be considered including reactor design. The conventional heterogeneous catalysis reactors include batch, continuous, and fluidized-bed reactors, while more recent setups include fixed-bed, microchannel, and multi-functional reactors. Other variables to consider are reactor dimensions, surface area, catalyst type, catalyst support, as well as reactor operating conditions such as temperature, pressure, and reactant concentrations.

Some large-scale industrial processes incorporating heterogeneous catalysts are listed below.

Process	Reactants, Product(s)	Catalyst	Comment
Sulfuric acid synthesis (Contact process)	$SO_2 + O_2, SO_3$	vanadium oxides	Hydration of SO_3 gives H_2SO_4
Ammonia synthesis (Haber–Bosch process)	$N_2 + H_2, NH_3$	iron oxides on alumina(Al_2O_3)	Consumes 1% of world's industrial energy budget
Nitric acid synthesis (Ostwald process)	$NH_3 + O_2, HNO_3$	unsupported Pt-Rh gauze	Direct routes from N_2 are uneconomical
Hydrogen production by Steam reforming	$CH_4 + H_2O, H_2 + CO_2$	Nickel or K_2O	Greener routes to H_2 by water splitting actively sought
Ethylene oxide synthesis	$C_2H_4 + O_2, C_2H_4O$	silver on alumina, with many promoters	Poorly applicable to other alkenes
Hydrogen cyanide synthesis (Andrussov oxidation)	$NH_3 + O_2 + CH_4, HCN$	Pt-Rh	Related ammoxidation process converts hydrocarbons to nitriles
Olefin polymerization Ziegler–Natta polymerization	propylene, polypropylene	$TiCl_3$ on $MgCl_2$	Many variations exist, including some homogeneous examples
Desulfurization of petroleum (hydrodesulfurization)	$H_2 + R_2S$ (idealized organosulfur impurity), $RH + H_2S$	Mo-Co on alumina	Produces low-sulfur hydrocarbons, sulfur recovered via the Claus process

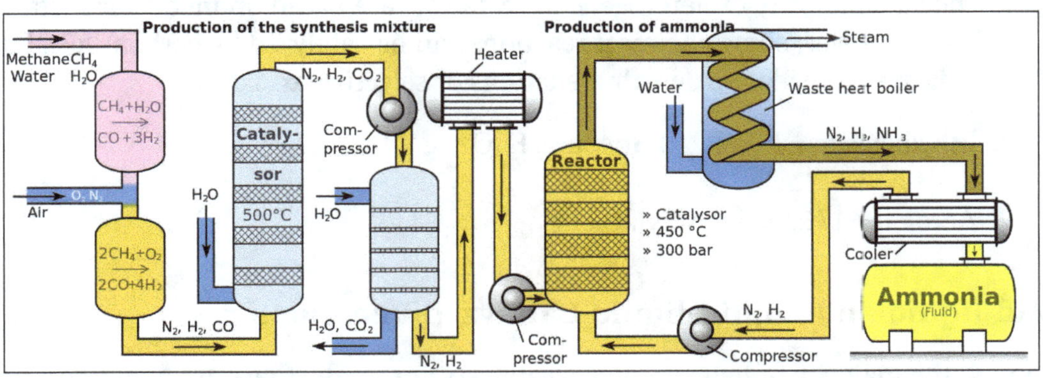

Process flow diagram illustrating the use of catalysis in the synthesis of ammonia (NH_3).

Other Examples

- Reduction of nitriles in the synthesis of phenethylamine with Raney nickel catalyst and hydrogen in ammonia:

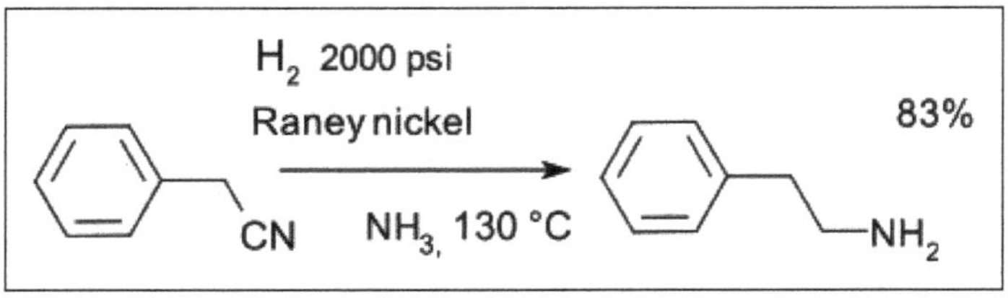

Nitrile hydrogenation.

- The cracking, isomerisation, and reformation of hydrocarbons to form appropriate and useful blends of petrol.
- In automobiles, catalytic converters are used to catalyze three main reactions:
    - The oxidation of carbon monoxide to carbon dioxide:

    $$2CO(g) + O_2(g) \rightarrow 2CO_2(g)$$

    - The reduction of nitrogen monoxide back to nitrogen:

    $$2NO(g) + 2CO(g) \rightarrow N_2(g) + 2CO_2(g)$$

    - The oxidation of hydrocarbons to water and carbon dioxide:

    $$2C_6H_6 + 15O_2 \rightarrow 12CO_2 + 6H_2O$$

- This process can occur with any of hydrocarbon, but most commonly is performed with petrol or diesel.
- Asymmetric heterogeneous catalysis facilitates the production of pure enantiomer compounds using chiral heterogeneous catalysts.
- The vast majority of heterogeneous catalysts are based on metals or metal oxides; however, some chemical reactions can be catalyzed by carbon-based materials, e.g., oxidative dehydrogenations or selective oxidations.

$$\text{Ethylbenzene} + 1/2\,O_2 \rightarrow \text{Styrene} + H_2O$$

$$\text{Acrolein} + 1/2\,O_2 \rightarrow \text{Acrylic acid}$$

Solid-liquid and Liquid-liquid Catalyzed Reactions

Although the majority of heterogeneous catalysts are solids, there are a few variations which are of practical value. For two immiscible solutions (liquids), one carries the catalyst while the other carries the reactant. This set up is the basis of biphasic catalysis as implemented in the industrial production of butyraldehyde by the hydroformylation of propylene.

Reacting phases	Examples given	Comment
solid + solution	hydrogenation of fatty acids with nickel	used for the production of margarine
immiscible liquid phases	hydroformylation of propene	aqueous phase catalyst; reactants and products mainly in non-aqueous phase

Homogeneous Catalysis

In chemistry, homogeneous catalysis is catalysis in a solution by a soluble catalyst. Homogeneous catalysis refers to catalytic reactions where the catalyst is in the same phase as the reactants. Homogeneous catalysis applies to reactions in the gas phase and even in solids. Heterogeneous catalysis is the alternative to homogeneous catalysis, where the catalysis occurs at the interface of two phases, typically gas-solid. The term is used almost exclusively to describe solutions and often implies catalysis by organometallic compounds.

Homogeneous catalysis using transition metal complexes is an area of research that has grown enormously in recent years. Many remarkable catalytic discoveries have been reported by researchers both in industry and in academia. The area is one of intense research and many practical applications, e.g., the production of acetic acid.

Enzymes are examples of homogeneous catalysts.

When materials react to form products, there may be a single reaction or multiple reactions occurring. A single reaction can be called an elementary. For most reactions, the sequence of elementary steps that constitute their mechanism is not known. Sometimes only a single reaction is observed. The reason is that the amount of the intermediate species formed is very small and escapes detection. The speed with which these intermediate species are created and destroyed also makes them difficult to detect. The general scheme is ‘reactants --> several intermediate species --> products’.In the case of a non-elementary reaction, the test of the validity of the mechanism proposed is whether the predicted kinetics corresponds to experimental observation. Sometimes reactions may proceed by more than one mechanism, or else their kinetics can be explained by several mechanisms. There are zero-order, first-order, second-order, and higher-order processes.

Advantages and Disadvantages

Advantages

- Homogeneous catalysts are effective at being highly selective towards producing the desired product.

- If a reaction is exothermic, it will release a lot of heat. It is easier to release heat from a solution (as one would do for a homogeneous catalyst) than if one were to use a heterogeneous catalyst, which tends to be an insoluble solid in the solution that adsorbs reaction participants onto it.
- Reactants can easily access the homogeneous catalyst because it is in solution already! This promotes high catalytic activity.
- Species in solution are easier to characterize (e.g. with various spectroscopy techniques) than species adsorbed onto a solid surface.

Disadvantages

- A catalyst capable of dissolving in solution will need to be separated later if it is to be recycled for re-usage (as per the principles of green chemistry).
- The homogeneous catalyst may have issues at high temperatures since it is not desirable for the solution (which contains the catalyst) and any volatile reactants to evaporate, even though high temperatures usually promote faster reactions.

Examples:

Acid Catalysis

The proton is the most pervasive homogeneous catalyst because water is the most common solvent. Water forms protons by the process of self-ionization of water. In an illustrative case, acids accelerate (catalyze) the hydrolysis of esters:

$$CH_3CO_2CH_3 + H_2O \rightleftharpoons CH_3CO_2H + CH_3OH$$

In the absence of acids, aqueous solutions of most esters do not hydrolyze at practical rates.

Transition Metal-catalysis

Many transformations classically considered homogeneous catalysis utilize soluble coordination and organometallic compounds as catalysts. These catalysts differ from solid catalysts, which are traditionally associated with heterogeneous catalysis. Homogeneous catalysis can be subdivided according to their substrates or their reductive/oxidative character.

Reductions

A prominent class of reductive transformations involve hydrogenation. H_2 is added to unsaturated substrates directly using hydrogen gas or indirectly by transferring hydrogen from substrates. The latter is called transfer hydrogenation. Related reactions entail "HX additions" where X = silyl (hydrosilylation) and CN (hydrocyanation).

Carbonylations

Hydroformylation is a prominent form of carbonylation, involving the addition of H and "C(O)H" across a double bond. MeOH and CO react in the presence of homogeneous catalysts to give acetic acid, as practiced in the Monsanto process and Cativa processes. Carbon monoxide is a common substrate in homogeneous catalysis and diverse reactions have been commercialized, including hydrocarboxylation and hydroesterifications.

Polymerization and Metathesis of Alkenes

A number of polyolefins are produced by Ziegler-Natta catalysis. Alkenes such as ethylene, propylene, and butadiene are common substrates. Olefin metathesis is usually practiced heterogeneously, but homogeneous variants are known.

Oxidations

Homogeneous catalysts are also used in a variety of oxidations. In the Wacker process, acetaldehyde is produced from ethylene to oxygen. Many non-organometallic complexes are also widely used in catalysis, e.g. for the production of terephthalic acid from xylene. Alkenes are epoxidized and dihydroxylated by metal complexes, as illustrated by the Halcon Process.

Hydration and Hydrolysis

Water is a common reagent in homogeneous catalysis. Esters and amides are slow to hydrolyze in neutral water, but the rates are sharply affected by metal complexes. The fastest and most selective catalysts for such reactions are often metalloenzymes, which can be viewed as large coordination complexes. The hydration of nitriles, alkenes, and alkynes are all catalyzed by metal complexes.

Other Forms of Homogeneous Catalysis

Enzymes are homogeneous catalysts that are essential for life but are also harnessed for industrial processes. A well-studied example is carbonic anhydrase, which catalyzes the release of CO_2 into the lungs from the bloodstream.

Industrial Applications

- Neohexene (3,3-dimethyl-1-butene): An important intermediate in the synthesis of fine chemicals, is produced from the dimer of isobutene, which consists of a mixture of 2,4,4-trimethyl-2-pentene and 2,4,4-trimethyl-1-pentene. Cross-metathesis of the former with ethene yields the desired product. The catalyst is a mixture of WO_3/SiO_2 for metathesis and MgO for isomerization at 370°C and 30 bar. The isobutene is recycled to the isobutene dimerization unit.

- Vestenamer 8012: A predominantly trans polymer of cyclooctene made with a WCl_6 based catalyst (Degussa-Hüls), used in blends and poly-norbornene made by a catalyst generated from $RuCl_3$ in t-BuOH in the presence of air (Norsorex by Atofina, a polymer with a very high molecular weight).
- Polydicyclopentadiene (DCPD): Dicyclopentadiene is the Diels-Alder adduct of cyclopentadiene, an abundant product from the cracker in the refinery. It contains a strained norbornene ring and a less strained cyclopentene ring. ROMP leads to opening of the norbornene ring, but some reaction of the cyclopentene ring also takes place, which leads to cross-linking. 4. For a recent update of industrial applications, many references exist.

Contrast with Heterogeneous Catalysis

Homogeneous catalysis differs from heterogeneous catalysis in that the catalyst is in a different phase than the reactants. One example of heterogeneous catalysis is the petrochemical alkylation process, where the liquid reactants are immiscible with a solution containing the catalyst. Heterogeneous catalysis offers the advantage that products are readily separated from the catalyst, and heterogeneous catalysts are often more stable and degrade much slower than homogeneous catalysts. However, heterogeneous catalysts are difficult to study, so their reaction mechanisms are often unknown.

Enzymes possess properties of both homogeneous and heterogeneous catalysts. As such, they are usually regarded as a third, separate category of catalyst.

Enzyme Kinetics

Dihydrofolate reductase from *E. coli* with its two substrates dihydrofolate (right) and NADPH (left), bound in the active site. The protein is shown as a ribbon diagram, with alpha helices in red, beta sheets in yellow and loops in blue. Generated from 7DFR.

Enzyme kinetics is the study of the chemical reactions that are catalysed by enzymes. In enzyme kinetics, the reaction rate is measured and the effects of varying the conditions of the reaction are investigated. Studying an enzyme's kinetics in this way can reveal the catalytic mechanism of this enzyme, its role in metabolism, how its activity is controlled, and how a drug or an agonist might inhibit the enzyme.

Enzymes are usually protein molecules that manipulate other molecules—the enzymes' substrates. These target molecules bind to an enzyme's active site and are transformed into products through a series of steps known as the enzymatic mechanism

$$E + S \rightleftarrows ES \rightleftarrows ES^* \rightleftarrows EP \rightleftarrows E + P$$

These mechanisms can be divided into single-substrate and multiple-substrate mechanisms. Kinetic studies on enzymes that only bind one substrate, such as triosephosphate isomerase, aim to measure the affinity with which the enzyme binds this substrate and the turnover rate. Some other examples of enzymes are phosphofructokinase and hexokinase, both of which are important for cellular respiration (glycolysis).

When enzymes bind multiple substrates, such as dihydrofolate reductase, enzyme kinetics can also show the sequence in which these substrates bind and the sequence in which products are released. An example of enzymes that bind a single substrate and release multiple products are proteases, which cleave one protein substrate into two polypeptide products. Others join two substrates together, such as DNA polymerase linking a nucleotide to DNA. Although these mechanisms are often a complex series of steps, there is typically one *rate-determining step* that determines the overall kinetics. This rate-determining step may be a chemical reaction or a conformational change of the enzyme or substrates, such as those involved in the release of products from the enzyme.

Knowledge of the enzyme's structure is helpful in interpreting kinetic data. For example, the structure can suggest how substrates and products bind during catalysis; what changes occur during the reaction; and even the role of particular amino acid residues in the mechanism. Some enzymes change shape significantly during the mechanism; in such cases, it is helpful to determine the enzyme structure with and without bound substrate analogues that do not undergo the enzymatic reaction.

Not all biological catalysts are protein enzymes; RNA-based catalysts such as ribozymes and ribosomes are essential to many cellular functions, such as RNA splicing and translation. The main difference between ribozymes and enzymes is that RNA catalysts are composed of nucleotides, whereas enzymes are composed of amino acids. Ribozymes also perform a more limited set of reactions, although their reaction mechanisms and kinetics can be analysed and classified by the same methods.

General Principles

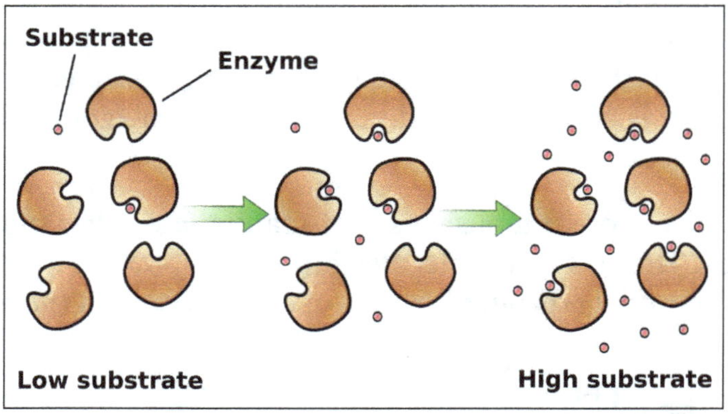

As larger amounts of substrate are added to a reaction, the available enzyme binding sites become filled to the limit of V_{Max}. Beyond this limit the enzyme is saturated with substrate and the reaction rate ceases to increase.

As larger amounts of substrate are added to a reaction, the available enzyme binding sites become filled to the limit of . Beyond this limit the enzyme is saturated with substrate and the reaction rate ceases to increase.

The reaction catalysed by an enzyme uses exactly the same reactants and produces exactly the same products as the uncatalysed reaction. Like other catalysts, enzymes do not alter the position of equilibrium between substrates and products. However, unlike uncatalysed chemical reactions, enzyme-catalysed reactions display saturation kinetics. For a given enzyme concentration and for relatively low substrate concentrations, the reaction rate increases linearly with substrate concentration; the enzyme molecules are largely free to catalyse the reaction, and increasing substrate concentration means an increasing rate at which the enzyme and substrate molecules encounter one another. However, at relatively high substrate concentrations, the reaction rate asymptotically approaches the theoretical maximum; the enzyme active sites are almost all occupied by substrates resulting in saturation, and the reaction rate is determined by the intrinsic turnover rate of the enzyme. The substrate concentration midway between these two limiting cases is denoted by K_M. Thus, K_M is the substrate concentration at which the reaction velocity is half of the maximum velocity.

The two most important kinetic properties of an enzyme are how easily the enzyme becomes saturated with a particular substrate, and the maximum rate it can achieve. Knowing these properties suggests what an enzyme might do in the cell and can show how the enzyme will respond to changes in these conditions.

Enzyme Assays

Progress curve for an enzyme reaction. The slope in the initial rate period is the initial rate of reaction v. The Michaelis–Menten equation describes how this slope varies with the concentration of substrate.

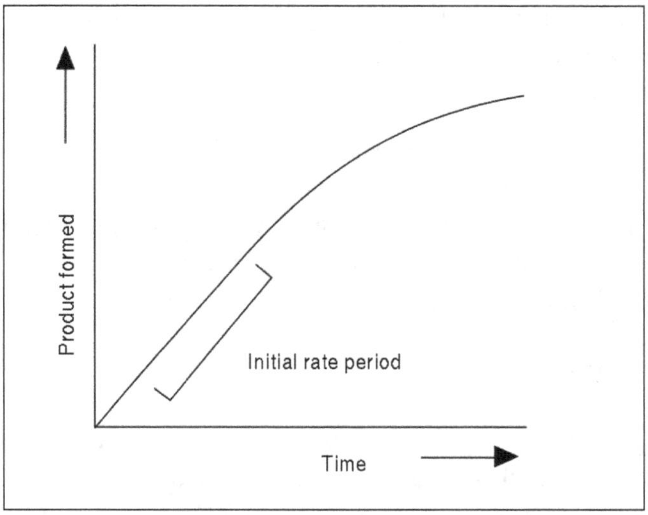

Enzyme assays are laboratory procedures that measure the rate of enzyme reactions. Since enzymes are not consumed by the reactions they catalyse, enzyme assays usually follow changes in the concentration of either substrates or products to measure the rate of reaction. There are many methods of measurement. Spectrophotometric assays observe change in the absorbance of light between products and reactants; radiometric assays involve the incorporation or release of radioactivity to measure the amount of product made over time. Spectrophotometric assays are most convenient since they allow the rate of the reaction to be measured continuously. Although radiometric assays require the removal and counting of samples (i.e., they are discontinuous assays) they are usually extremely sensitive and can measure very low levels of enzyme activity. An analogous approach is to use mass spectrometry to monitor the incorporation or release of stable isotopes as substrate is converted into product.

The most sensitive enzyme assays use lasers focused through a microscope to observe changes in single enzyme molecules as they catalyse their reactions. These measurements either use changes in the fluorescence of cofactors during an enzyme's reaction mechanism, or of fluorescent dyes added onto specific sites of the protein to report movements that occur during catalysis. These studies are providing a new view of the kinetics and dynamics of single enzymes, as opposed to traditional enzyme kinetics, which observes the average behaviour of populations of millions of enzyme molecules.

An example progress curve for an enzyme assay is shown above. The enzyme produces product at an initial rate that is approximately linear for a short period after the start of the reaction. As the reaction proceeds and substrate is consumed, the rate continuously slows (so long as substrate is not still at saturating levels). To measure the initial (and maximal) rate, enzyme assays are typically carried out while the reaction has progressed only a few percent towards total completion. The length of the initial rate period depends on the assay conditions and can range from milliseconds to hours.

However, equipment for rapidly mixing liquids allows fast kinetic measurements on initial rates of less than one second. These very rapid assays are essential for measuring pre-steady-state kinetics.

Most enzyme kinetics studies concentrate on this initial, approximately linear part of enzyme reactions. However, it is also possible to measure the complete reaction curve and fit this data to a non-linear rate equation. This way of measuring enzyme reactions is called progress-curve analysis. This approach is useful as an alternative to rapid kinetics when the initial rate is too fast to measure accurately.

Single-substrate Reactions

Enzymes with single-substrate mechanisms include isomerases such as triosephosphateisomerase or bisphosphoglycerate mutase, intramolecular lyases such as adenylate cyclase and the hammerhead ribozyme, an RNA lyase. However, some enzymes that only have a single substrate do not fall into this category of mechanisms. Catalase is an example of this, as the enzyme reacts with a first molecule of hydrogen peroxide substrate, becomes oxidised and is then reduced by a second molecule of substrate. Although a single substrate is involved, the existence of a modified enzyme intermediate means that the mechanism of catalase is actually a ping–pong mechanism.

Michaelis–Menten Kinetics

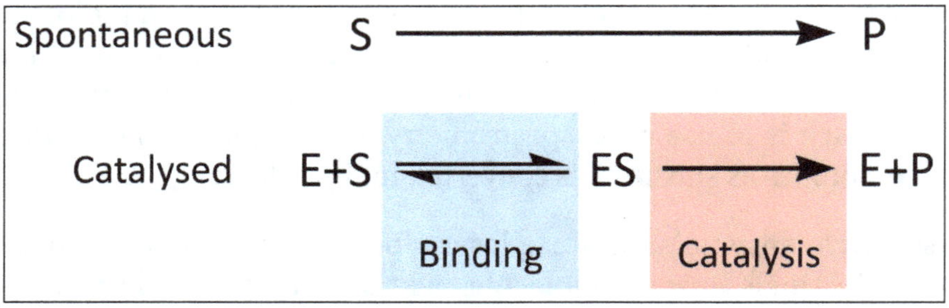

A chemical reaction mechanism with or without enzyme catalysis. The enzyme (E) binds substrate (S) to produce product (P).

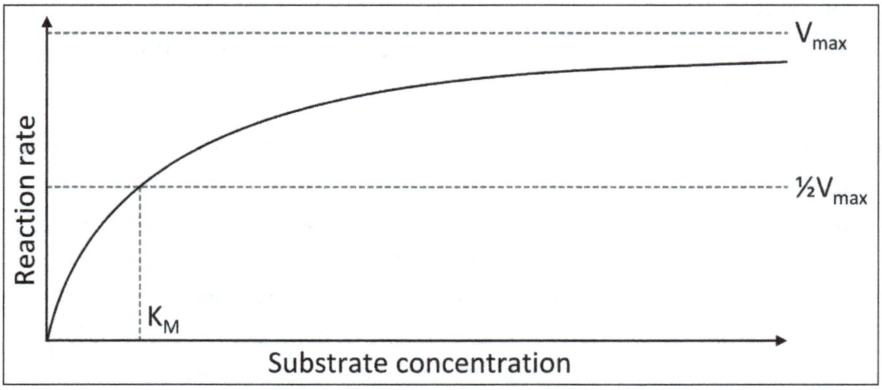

Saturation curve for an enzyme reaction showing the relation between the substrate concentration and reaction rate.

As enzyme-catalysed reactions are saturable, their rate of catalysis does not show a linear response to increasing substrate. If the initial rate of the reaction is measured over a range of substrate concentrations (denoted as [S]), the initial reaction rate (v_0) increases as [S] increases, as shown on the right. However, as [S] gets higher, the enzyme becomes saturated with substrate and the initial rate reaches V_{max}, the enzyme's maximum rate.

The Michaelis–Menten kinetic model of a single-substrate reaction is shown on the right. There is an initial bimolecular reaction between the enzyme E and substrate S to form the enzyme–substrate complex ES. The rate of enzymatic reaction increases with the increase of the substrate concentration up to a certain level called V_{max}; at V_{max}, increase in substrate concentration does not cause any increase in reaction rate as there is no more enzyme (E) available for reacting with substrate (S). Here, the rate of reaction becomes dependent on the ES complex and the reaction becomes a unimolecular reaction with an order of zero. Though the enzymatic mechanism for the unimolecular reaction $ES \xrightarrow{k_{cat}} E + P$ can be quite complex, there is typically one rate-determining enzymatic step that allows this reaction to be modelled as a single catalytic step with an apparent unimolecular rate constant k_{cat}. If the reaction path proceeds over one or several intermediates, k_{cat} will be a function of several elementary rate constants, whereas in the simplest case of a single elementary reaction (e.g. no intermediates) it will be identical to the elementary unimolecular rate constant k_2. The apparent unimolecular rate constant k_{cat} is also called *turnover number* and denotes the maximum number of enzymatic reactions catalysed per second.

The Michaelis–Menten equation describes how the (initial) reaction rate v_0 depends on the position of the substrate-binding equilibrium and the rate constant k_2.

$$v_0 = \frac{V_{max}[S]}{K_M + [S]} \quad \textit{(Michaelis–Menten equation)}$$

with the constants,

$$K_M \stackrel{\text{def}}{=} \frac{k_2 + k_{-1}}{k_1} \approx K_D$$

$$V_{max} \stackrel{\text{def}}{=} k_{cat}[E]_{tot}$$

This Michaelis–Menten equation is the basis for most single-substrate enzyme kinetics. Two crucial assumptions underlie this equation (apart from the general assumption about the mechanism only involving no intermediate or product inhibition, and there is no allostericity or cooperativity). The first assumption is the so-called quasi-steady-state assumption (or pseudo-steady-state hypothesis), namely that the concentration of the substrate-bound enzyme (and hence also the unbound enzyme) changes much more slowly than those of the product and substrate and thus the change over time of the complex can be set to zero $d[\mathrm{ES}]/dt \overset{!}{=} 0$. The second assumption is that the total enzyme concentration does not change over time, thus $[\mathrm{E}]_{\mathrm{tot}} = [\mathrm{E}] + [\mathrm{ES}] \overset{!}{=} \mathrm{const}$.

The Michaelis constant K_M is experimentally defined as the concentration at which the rate of the enzyme reaction is half V_{max}, which can be verified by substituting [S] = K_M into the Michaelis–Menten equation and can also be seen graphically. If the rate-determining enzymatic step is slow compared to substrate dissociation ($k_2 \ll k_{-1}$), the Michaelis constant K_M is roughly the dissociation constant K_D of the ES complex.

If [S] is small compared to K_M then the term $[\mathrm{S}]/(K_M + [\mathrm{S}]) \approx [\mathrm{S}]/K_M$ and also very little ES complex is formed, thus $[\mathrm{E}]_{\mathrm{tot}} \approx [\mathrm{E}]$. Therefore, the rate of product formation is,

$$v_0 \approx \frac{k_{cat}}{K_M}[\mathrm{E}][\mathrm{S}] \qquad \text{if } [\mathrm{S}] \ll K_M$$

Thus the product formation rate depends on the enzyme concentration as well as on the substrate concentration, the equation resembles a bimolecular reaction with a corresponding pseudo-second order rate constant k_2 / K_M. This constant is a measure of catalytic efficiency. The most efficient enzymes reach a $k \quad K$ in the range of 10^8 – 10^{10} $\mathrm{M}^{-1}\,\mathrm{s}^{-1}$. These enzymes are so efficient they effectively catalyse a reaction each time they encounter a substrate molecule and have thus reached an upper theoretical limit for efficiency (diffusion limit); and are sometimes referred to as kinetically perfect enzymes. But most enzymes are far from perfect: the average values of k_2 / K_M and k_2 are about $10^5\,\mathrm{s}^{-1}\mathrm{M}^{-1}$ and $10\,\mathrm{s}^{-1}$, respectively.

Direct use of the Michaelis–Menten Equation for Time Course Kinetic Analysis

The observed velocities predicted by the Michaelis–Menten equation can be used to directly model the time course disappearance of substrate and the production of product through incorporation of the Michaelis–Menten equation into the equation for first order chemical kinetics. This can only be achieved however if one recognises the problem associated with the use of Euler's number in the description of first order chemical kinetics. i.e. e^{-k} is a split constant that introduces a systematic error into calculations and can be rewritten as a single constant which represents the remaining substrate after each time period.

$$[S] = [S]_0(1-k)^t$$

$$[S] = [S]_0(1-v/[S]_0)^t$$

$$[S] = [S]_0(1-(V_{max}[S]_0/(K_M+[S]_0)/[S]_0))^t$$

In 1983 Stuart Beal (and also independently Santiago Schnell and Claudio Mendoza in 1997) derived a closed form solution for the time course kinetics analysis of the Michaelis-Menten mechanism. The solution, known as the Schnell-Mendoza equation, has the form:

$$\frac{[S]}{K_M} = W[F(t)]$$

where W[ ] is the Lambert-W function. and where F(t) is,

$$F(t) = \frac{[S]_0}{K_M} \exp\left(\frac{[S]_0}{K_M} - \frac{V_{max}}{K_M} t\right)$$

This equation is encompassed by the equation below, obtained by Berberan-Santos, which is also valid when the initial substrate concentration is close to that of enzyme,

$$\frac{[S]}{K_M} = W\left[F(t)\right] - \frac{V_{max}}{k_{cat} K_M} \frac{W\left[F(t)\right]}{1 + W\left[F(t)\right]}$$

where W[ ] is again the Lambert-W function.

Linear Plots of the Michaelis–Menten Equation

Lineweaver–Burk or double-reciprocal plot of kinetic data, showing the significance of the axis intercepts and gradient.

The plot of v versus [S] above is not linear; although initially linear at low [S], it bends over to saturate at high [S]. Before the modern era of nonlinear curve-fitting on computers, this nonlinearity could make it difficult to estimate K_M and V_{max} accurately. Therefore, several researchers developed linearisations of the Michaelis–Menten equation, such as the Lineweaver–Burk plot, the Eadie–Hofstee diagram and the Hanes–Woolf plot. All of these linear representations can be useful for visualising data, but none should be used to determine kinetic parameters, as computer software is readily available that allows for more accurate determination by nonlinear regression methods.

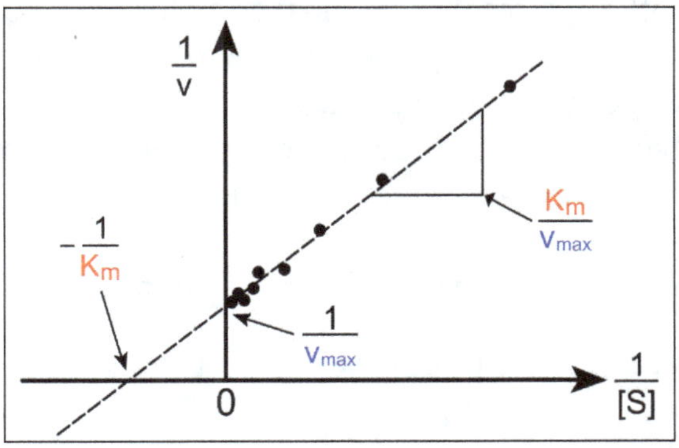

The Lineweaver–Burk plot or double reciprocal plot is a common way of illustrating kinetic data. This is produced by taking the reciprocal of both sides of the Michaelis–Menten equation. As shown on the right, this is a linear form of the Michaelis–Menten equation and produces a straight line with the equation y = mx + c with a y-intercept equivalent to $1/V_{max}$ and an x-intercept of the graph representing $-1/K_M$.

$$\frac{1}{v} = \frac{K_M}{V_{max}[S]} + \frac{1}{V_{max}}$$

Naturally, no experimental values can be taken at negative 1/[S]; the lower limiting value 1/[S] = 0 (the y-intercept) corresponds to an infinite substrate concentration, where $1/v = 1/V_{max}$ as shown at the right; thus, the x-intercept is an extrapolation of the experimental data taken at positive concentrations. More generally, the Lineweaver–Burk plot skews the importance of measurements taken at low substrate concentrations and, thus, can yield inaccurate estimates of V_{max} and K_M. A more accurate linear plotting method is the Eadie–Hofstee plot. In this case, v is plotted against v/[S]. In the third common linear representation, the Hanes–Woolf plot, [S]/v is plotted against [S]. In general, data normalisation can help diminish the amount of experimental work and can increase the reliability of the output, and is suitable for both graphical and numerical analysis.

Practical Significance of Kinetic Constants

The study of enzyme kinetics is important for two basic reasons. Firstly, it helps explain how enzymes work, and secondly, it helps predict how enzymes behave in living organisms. The kinetic constants defined above, K_M and V_{max}, are critical to attempts to understand how enzymes work together to control metabolism.

Making these predictions is not trivial, even for simple systems. For example, oxaloacetate is formed by malate dehydrogenase within the mitochondrion. Oxaloacetate can then be consumed by citrate synthase, phosphoenolpyruvate carboxykinase or aspartate aminotransferase, feeding into the citric acid cycle, gluconeogenesis or aspartic acid biosynthesis, respectively. Being able to predict how much oxaloacetate goes into which pathway requires knowledge of the concentration of oxaloacetate as well as the concentration and kinetics of each of these enzymes. This aim of predicting the behaviour of metabolic pathways reaches its most complex expression in the synthesis of huge amounts of kinetic and gene expression data into mathematical models of entire organisms. Alternatively, one useful simplification of the metabolic modelling problem is to ignore the underlying enzyme kinetics and only rely on information about the reaction network's stoichiometry, a technique called flux balance analysis.

Michaelis–Menten Kinetics with Intermediate

One could also consider the less simple case:

$$E + S \underset{k_{-1}}{\overset{k_1}{\rightleftharpoons}} ES \xrightarrow{k_2} EI \xrightarrow{k_3} E + P$$

where a complex with the enzyme and an intermediate exists and the intermediate is converted into product in a second step. In this case we have a very similar equation

$$v_0 = k_{cat} \frac{[S][E]_0}{K_M^{'} + [S]}$$

but the constants are different,

$$K_M^{'} \stackrel{\text{def}}{=} \frac{k_3}{k_2 + k_3} K_M = \frac{k_3}{k_2 + k_3} \cdot \frac{k_2 + k_{-1}}{k_1}$$

$$k_{cat} \stackrel{\text{def}}{=} \frac{k_3 k_2}{k_2 + k_3}$$

We see that for the limiting case $k_3 \gg k_2$, thus when the last step from $EI \rightarrow E + P$ is much faster than the previous step, we get again the original equation. Mathematically we have then $K_M^{'} \approx K_M$ and $k_{cat} \approx k_2$.

Multi-substrate Reactions

Multi-substrate reactions follow complex rate equations that describe how the substrates bind and in what sequence. The analysis of these reactions is much simpler if the concentration of substrate A is kept constant and substrate B varied. Under these conditions, the enzyme behaves just like a single-substrate enzyme and a plot of v by [S] gives apparent K_M and V_{max} constants for substrate B. If a set of these measurements is performed at different fixed concentrations of A, these data can be used to work out what the mechanism of the reaction is. For an enzyme that takes two substrates A and B and turns them into two products P and Q, there are two types of mechanism: ternary complex and ping–pong.

Ternary-complex Mechanisms

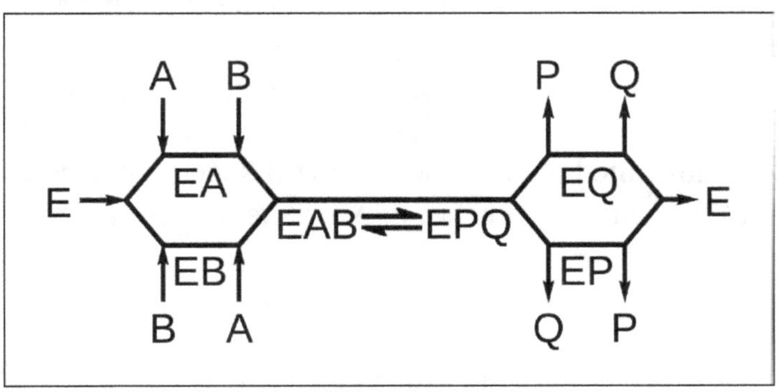

Random-order ternary-complex mechanism for an enzyme reaction. The reaction path is shown as a line and enzyme intermediates containing substrates A and B or products P and Q are written below the line.

In these enzymes, both substrates bind to the enzyme at the same time to produce an EAB ternary complex. The order of binding can either be random (in a random mechanism) or substrates have to bind in a particular sequence (in an ordered mechanism). When a set of v by [S] curves (fixed A, varying B) from an enzyme with a ternary-complex mechanism are plotted in a Lineweaver–Burk plot, the set of lines produced will intersect.

Enzymes with ternary-complex mechanisms include glutathione *S*-transferase, dihydrofolate reductase and DNA polymerase. The following links show short animations of the ternary-complex mechanisms of the enzymes dihydrofolate reductase[β] and DNA polymerase[γ].

Ping–pong Mechanisms

$$E \rightarrow [\xrightarrow{\overset{A}{\downarrow}}]EA \Leftrightarrow E^{*}P \rightarrow [\xrightarrow{\overset{P}{\uparrow}}]E^{*} \rightarrow [\xrightarrow{\overset{B}{\downarrow}}]E^{*}B \Leftrightarrow EQ \rightarrow [\xrightarrow{\overset{Q}{\uparrow}}]E$$

Ping–pong mechanism for an enzyme reaction. Intermediates contain substrates A and B or products P and Q.

As shown on the right, enzymes with a ping-pong mechanism can exist in two states, E and a chemically modified form of the enzyme E*; this modified enzyme is known as an intermediate. In such mechanisms, substrate A binds, changes the enzyme to E* by, for example, transferring a chemical group to the active site, and is then released. Only after the first substrate is released can substrate B bind and react with the modified enzyme, regenerating the unmodified E form. When a set of v by [S] curves (fixed A, varying B) from an enzyme with a ping–pong mechanism are plotted in a Lineweaver–Burk plot, a set of parallel lines will be produced. This is called a secondary plot.

Enzymes with ping–pong mechanisms include some oxidoreductases such as thioredoxin peroxidase, transferases such as acylneuraminate cytidylyltransferase and serine proteases such as trypsin and chymotrypsin. Serine proteases are a very common and diverse family of enzymes, including digestive enzymes (trypsin, chymotrypsin, and elastase), several enzymes of the blood clotting cascade and many others. In these serine proteases, the E* intermediate is an acyl-enzyme species formed by the attack of an active site serine residue on a peptide bond in a protein substrate.

Reversible Catalysis and the Haldane Equation

We consider the case of an enzyme that can catalyse the reaction in both directions (not all enzymes can):

$$E + S \underset{k_{-1}}{\overset{k_1}{\rightleftharpoons}} ES \underset{k_{-2}}{\overset{k_2}{\rightleftharpoons}} E + P$$

The steady-state, initial rate of the reaction is $v_0 = \frac{d[\mathrm{P}]}{dt} = \frac{(k_1 k_2[\mathrm{S}] - k_{-1}k_{-2}[\mathrm{P}])[\mathrm{E}]_0}{k_{-1} + k_2 + k_1[\mathrm{S}] + k_{-2}[\mathrm{P}]}$

v_0 is positive if the reaction proceed in the forward direction ($S \rightarrow P$) and negative otherwise.

Equilibrium requires that $v = 0$, which occurs when $\frac{[\mathrm{P}]_{\mathrm{eq}}}{[\mathrm{S}]_{\mathrm{eq}}} = \frac{k_1 k_2}{k_{-1}k_{-2}} = K_{\mathrm{eq}}$. This shows that thermodynamics forces a relation between the values of the 4 rate constants.

The values of the forward and backward *maximal* rates, obtained for $[\mathrm{S}] \rightarrow \infty$, $[\mathrm{P}] = 0$, and $[\mathrm{S}] = 0$, $[\mathrm{P}] \rightarrow \infty$, respectively, are $V^f_{\max} = k_2[\mathrm{E}]_{tot}$ and $V^b_{\max} = -k_{-1}[\mathrm{E}]_{tot}$, respectively. Their ratio is not equal to the equilibrium constant, which implies that thermodynamics does not constrain the ratio of the maximal rates. This explains that enzymes can be much "better catalysts" (*in terms of maximal rates*) in one particular direction of the reaction.

On can also derive the two Michaelis constants $K^S_M = (k_{-1} + k_2) / k_1 a$ and $K^P_M = (k_{-1} + k_2) / k_{-2}$. The Haldane equation is the relation $K_{\mathrm{eq}} = \frac{[\mathrm{P}]_{\mathrm{eq}}}{[\mathrm{S}]_{\mathrm{eq}}} = \frac{V^f_{\max} / K^S_M}{V^b_{\max} / K^P_M}$.

Therefore, thermodynamics constrains the ratio between the forward and backward $V_{\max} / K_M$ values, not the ratio of $V_{\max}$ values.

Non-Michaelis–Menten Kinetics

Some enzymes produce a sigmoid v by [S] plot, which often indicates cooperative binding of substrate to the active site. This means that the binding of one substrate molecule affects the binding of subsequent substrate molecules. This behavior is most common in multimeric enzymes with several interacting active sites. Here, the mechanism of cooperation is similar to that of hemoglobin, with binding of substrate to one active site altering the affinity of the other active sites for substrate molecules. Positive cooperativity occurs when binding of the first substrate molecule *increases* the affinity of the other active sites for substrate. Negative cooperativity occurs when binding of the first substrate *decreases* the affinity of the enzyme for other substrate molecules.

Allosteric enzymes include mammalian tyrosyl tRNA-synthetase, which shows negative cooperativity, and bacterial aspartate transcarbamoylase and phosphofructokinase, which show positive cooperativity.

Cooperativity is surprisingly common and can help regulate the responses of enzymes to changes in the concentrations of their substrates. Positive cooperativity makes enzymes much more sensitive to [S] and their activities can show large changes over a narrow range of substrate concentration. Conversely, negative cooperativity makes enzymes insensitive to small changes in [S].

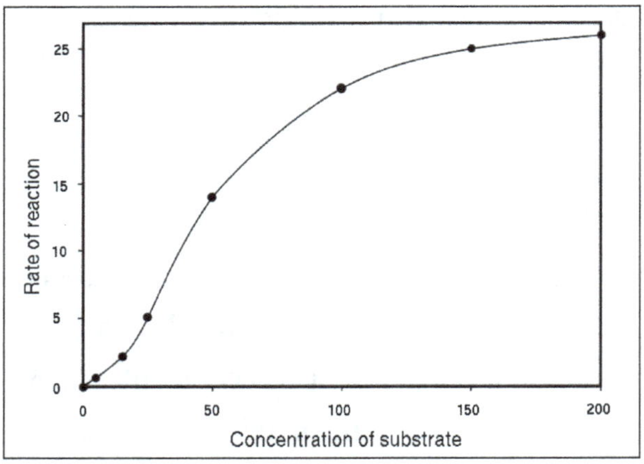

Saturation curve for an enzyme reaction showing sigmoid kinetics.

The Hill equation (biochemistry) is often used to describe the degree of cooperativity quantitatively in non-Michaelis–Menten kinetics. The derived Hill coefficient n measures how much the binding of substrate to one active site affects the binding of substrate to the other active sites. A Hill coefficient of <1 indicates negative cooperativity and a coefficient of >1 indicates positive cooperativity.

Pre-steady-state Kinetics

In the first moment after an enzyme is mixed with substrate, no product has been formed and no intermediates exist. The study of the next few milliseconds of the reaction is called pre-steady-state kinetics. Pre-steady-state kinetics is therefore concerned with the formation and consumption of enzyme–substrate intermediates (such as ES or E*) until their steady-state concentrations are reached.

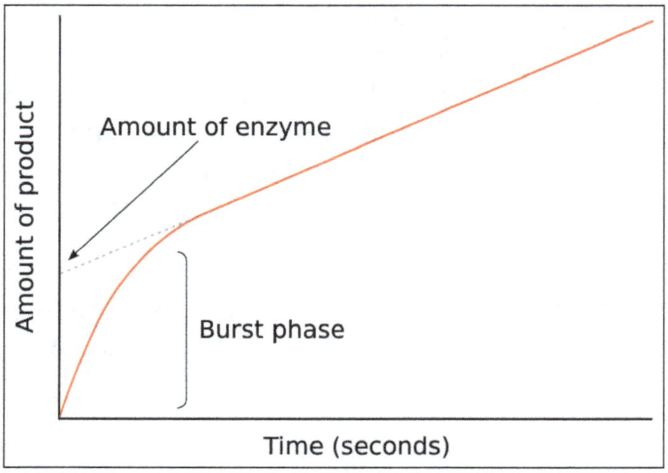

Pre-steady state progress curve, showing the burst phase of an enzyme reaction.

This approach was first applied to the hydrolysis reaction catalysed by chymotrypsin. Often, the detection of an intermediate is a vital piece of evidence in investigations of what mechanism an enzyme follows. For example, in the ping–pong mechanisms that are shown, rapid kinetic measurements can follow the release of product P and measure the formation of the modified enzyme intermediate E*. In the case of chymotrypsin, this intermediate is formed by an attack on the substrate by the nucleophilic serine in the active site and the formation of the acyl-enzyme intermediate.

In the figure to the right, the enzyme produces E* rapidly in the first few seconds of the reaction. The rate then slows as steady state is reached. This rapid burst phase of the reaction measures a single turnover of the enzyme. Consequently, the amount of product released in this burst, shown as the intercept on the *y*-axis of the graph, also gives the amount of functional enzyme which is present in the assay.

Chemical Mechanism

An important goal of measuring enzyme kinetics is to determine the chemical mechanism of an enzyme reaction, i.e., the sequence of chemical steps that transform substrate into product. The kinetic approaches discussed above will show at what rates intermediates are formed and inter-converted, but they cannot identify exactly what these intermediates are.

Kinetic measurements taken under various solution conditions or on slightly modified enzymes or substrates often shed light on this chemical mechanism, as they reveal the rate-determining step or intermediates in the reaction. For example, the breaking of a covalent bond to a hydrogen atom is a common rate-determining step. Which of the possible hydrogen transfers is rate determining can be shown by measuring the kinetic effects of substituting each hydrogen by deuterium, its stable isotope. The rate will change when the critical hydrogen is replaced, due to a primary kinetic isotope effect, which occurs because bonds to deuterium are harder to break than bonds to hydrogen. It is also possible to measure similar effects with other isotope substitutions, such as $^{13}C/^{12}C$ and $^{18}O/^{16}O$, but these effects are more subtle.

Isotopes can also be used to reveal the fate of various parts of the substrate molecules in the final products. For example, it is sometimes difficult to discern the origin of an oxygen atom in the final product; since it may have come from water or from part of the substrate. This may be determined by systematically substituting oxygen's stable isotope ^{18}O into the various molecules that participate in the reaction and checking for the isotope in the product. The chemical mechanism can also be elucidated by examining the kinetics and isotope effects under different pH conditions, by altering the metal ions or other bound cofactors, by site-directed mutagenesis of conserved amino acid residues, or by studying the behaviour of the enzyme in the presence of analogues of the substrates.

Enzyme Inhibition and Activation

Enzyme inhibitors are molecules that reduce or abolish enzyme activity, while enzyme activators are molecules that increase the catalytic rate of enzymes. These interactions can be either *reversible* (i.e., removal of the inhibitor restores enzyme activity) or *irreversible* (i.e., the inhibitor permanently inactivates the enzyme).

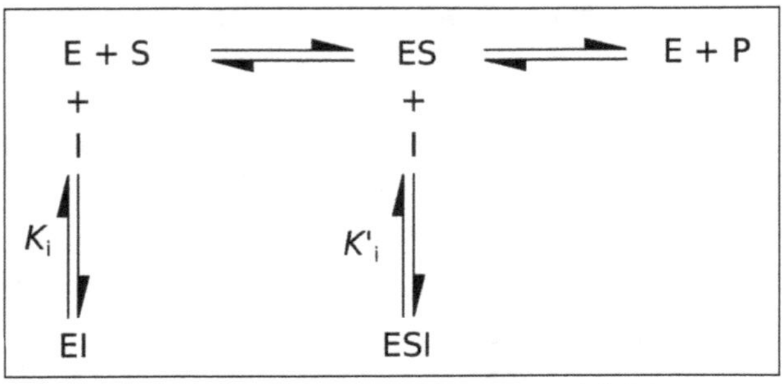

Kinetic scheme for reversible enzyme inhibitors.

Reversible Inhibitors

Traditionally reversible enzyme inhibitors have been classified as competitive, uncompetitive, or non-competitive, according to their effects on K_M and V_{max}. These different effects result from the inhibitor binding to the enzyme E, to the enzyme–substrate complex ES, or to both, respectively. The division of these classes arises from a problem in their derivation and results in the need to use two different binding constants for one binding event. The binding of an inhibitor and its effect on the enzymatic activity are two distinctly different things, another problem the traditional equations fail to acknowledge. In noncompetitive inhibition the binding of the inhibitor results in 100% inhibition of the enzyme only, and fails to consider the possibility of anything in between. In noncompetitive inhibition, the inhibitor will bind to an enzyme at its allosteric site; therefore, the binding affinity, or inverse of K_M, of the substrate with the enzyme will remain the same. On the other hand, the V_{max} will decrease relative to an uninhibited enzyme. On a Lineweaver-Burk plot, the presence of a noncompetitive inhibitor is illustrated by a change in the y-intercept, defined as $1/V_{max}$. The x-intercept, defined as $-1/K_M$, will remain the same. In competitive inhibition, the inhibitor will bind to an enzyme at the active site, competing with the substrate. As a result, the K_M will increase and the V_{max} will remain the same. The common form of the inhibitory term also obscures the relationship between the inhibitor binding to the enzyme and its relationship to any other binding term be it the Michaelis–Menten equation or a dose response curve associated with ligand receptor binding. To demonstrate the relationship the following rearrangement can be made:

$$\frac{V_{max}}{1+\frac{[I]}{K_i}} = \frac{V_{max}}{\frac{[I]+K_i}{K_i}}$$

Adding zero to the bottom ([I]-[I]):

$$\frac{V_{\max}}{\frac{[I]+K_i}{[I]+K_i-[I]}}$$

Dividing by [I]+K_i,

$$\frac{V_{\max}}{\frac{1}{1-\frac{[I]}{[I]+K_i}}}=V_{\max}-V_{\max}\frac{[I]}{[I]+K_i}$$

This notation demonstrates that similar to the Michaelis–Menten equation, where the rate of reaction depends on the percent of the enzyme population interacting with substrate, the effect of the inhibitor is a result of the percent of the enzyme population interacting with inhibitor. The only problem with this equation in its present form is that it assumes absolute inhibition of the enzyme with inhibitor binding, when in fact there can be a wide range of effects anywhere from 100% inhibition of substrate turn over to just >0%. To account for this the equation can be easily modified to allow for different degrees of inhibition by including a delta $V_{\max}$ term.

$$V_{\max}-\Delta V_{\max}\frac{[I]}{[I]+K_i}$$

or

$$V_{\max 1}-(V_{\max 1}-V_{\max 2})\frac{[I]}{[I]+K_i}$$

This term can then define the residual enzymatic activity present when the inhibitor is interacting with individual enzymes in the population. However the inclusion of this term has the added value of allowing for the possibility of activation if the secondary $V_{\max}$ term turns out to be higher than the initial term. To account for the possibly of activation as well the notation can then be rewritten replacing the inhibitor "I" with a modifier term denoted here as "X".

$$V_{\max 1}-(V_{\max 1}-V_{\max 2})\frac{[X]}{[X]+K_x}$$

While this terminology results in a simplified way of dealing with kinetic effects relating to the maximum velocity of the Michaelis–Menten equation, it highlights potential problems with the term used to describe effects relating to the K_M. The K_M relating to the affinity of the enzyme for the substrate should in most cases relate to potential changes in the binding site of the enzyme which would directly result from enzyme inhibitor interactions. As such a term similar to the one proposed above to modulate V_{max} should be appropriate in most situations:

$$K_{m1} - (K_{m1} - K_{m2})\frac{[X]}{[X] + K_x}$$

Irreversible Inhibitors

Enzyme inhibitors can also irreversibly inactivate enzymes, usually by covalently modifying active site residues. These reactions, which may be called suicide substrates, follow exponential decay functions and are usually saturable. Below saturation, they follow first order kinetics with respect to inhibitor. Irreversible inhibition could be classified into two distinct types. Affinity labelling is a type of irreversible inhibition where a functional group that is highly reactive modifies a catalytically critical residue on the protein of interest to bring about inhibition. Mechanism-based inhibition, on the other hand, involves binding of the inhibitor followed by enzyme mediated alterations that transform the latter into a reactive group that irreversibly modifies the enzyme.

Mechanisms of Catalysis

The energy variation as a function of reaction coordinate shows the stabilisation of the transition state by an enzyme.

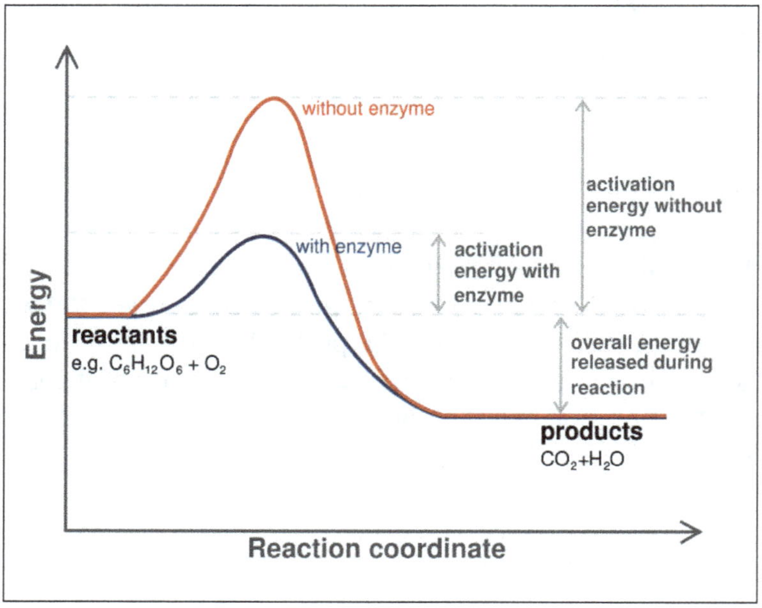

The favoured model for the enzyme–substrate interaction is the induced fit model. This model proposes that the initial interaction between enzyme and substrate is relatively weak, but that these weak interactions rapidly induce conformational changes in the enzyme that strengthen binding. These conformational changes also bring catalytic residues in the active site close to the chemical bonds in the substrate that will be altered in the reaction. Conformational changes can be measured using circular dichroism or dual polarisation interferometry. After binding takes place, one or more mechanisms of catalysis lower the energy of the reaction's transition state by providing an alternative chemical pathway for the reaction. Mechanisms of catalysis include catalysis by bond strain; by proximity and orientation; by active-site proton donors or acceptors; covalent catalysis and quantum tunnelling.

Enzyme kinetics cannot prove which modes of catalysis are used by an enzyme. However, some kinetic data can suggest possibilities to be examined by other techniques. For example, a ping–pong mechanism with burst-phase pre-steady-state kinetics would suggest covalent catalysis might be important in this enzyme's mechanism. Alternatively, the observation of a strong pH effect on V_{max} but not K_M might indicate that a residue in the active site needs to be in a particular ionisation state for catalysis to occur.

Enzyme Activator

Enzyme activators are molecules that bind to enzymes and increase their activity. They are the opposite of enzyme inhibitors. These molecules are often involved in the allosteric regulation of enzymes in the control of metabolism. An example of an enzyme activator working in this way is fructose 2,6-bisphosphate, which activates phosphofructokinase 1 and increases the rate of glycolysis in response to the hormone insulin. In some cases, when a substrate binds to one catalytic subunit of an enzyme, this can trigger an increase in the substrate affinity as well as catalytic activity in the enzyme's other subunits, and thus the substrate acts as an activator.

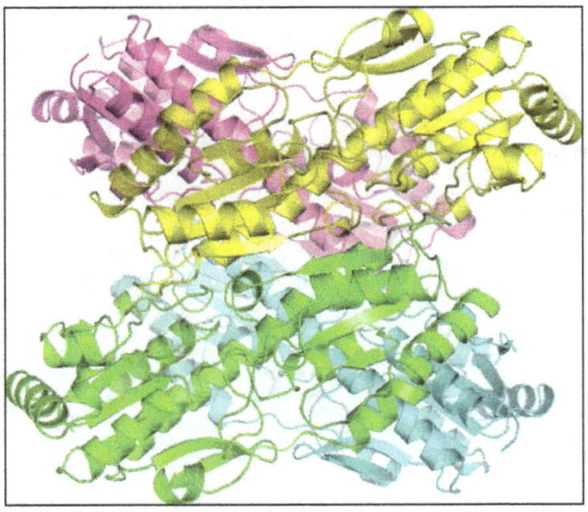

Bacillus stearothermophilus phosphofructokinase.

Examples:

Hexokinase-I

Hexokinase-I (HK-I) is an enzyme activator because it draws glucose into the glycolysis pathway. Its function is to phosphorylate glucose releasing glucose-6-phosphate (G6P) as the product. HK-I not only signals the activation of glucose into glycolysis but also maintains a low glucose concentration to facilitate glucose diffusion into the cell. It has two catalytic domains (N-terminal domain and C-terminal domain) which are connected through an α-helix. The N-terminal acts as an allosteric regulator of C-terminal; the C-terminal is the only one involved in the catalytic activity. HK-I is regulated by the concentration of G6P, where G6P acts as a feedback inhibitor. At low G6P concentration, HK-I is activated; at high G6P concentration, the HK-I is inhibited.

Glucokinase

Glucokinase (GK) is an enzyme that helps in the glycolytic pathway by phosphorylating glucose into glucose-6-phosphate (G6P). It is an isozyme of hexokinase and is found mainly in pancreatic β cells, but also liver, gut, and brain cells where glycolysis cause glucose-induced insulin secretion. Glucokinase activator lowers blood glucose concentrations by enhancing glucose uptake in the liver and increasing insulin production by the pancreatic β cells. Due to this, Glucokinase and glucokinase activators are the focus of treatment for those with type 2 diabetes mellitus. Glucokinase have a single allosteric site where the glucose-regulating protein (GKRP) binds in the nucleus of the cell in its inactive form when there is a low concentration of glucose present in the cell. However, when the glucose concentration of the cell increases the glucokinase-GKRP complex breaks apart and GK proceeds to the cytoplasm where it then phosphorylates glucose. Glucose when abundant in cells acts as an enzyme activator for glucokinase. Glucokinase activation in the β cells and liver cells results in the uptake of glucose and production of glycogen. This activation in the β cells leads to insulin secretion, promoting glucose uptake storing it as glycogen in the muscles.

Enzyme Inhibitor

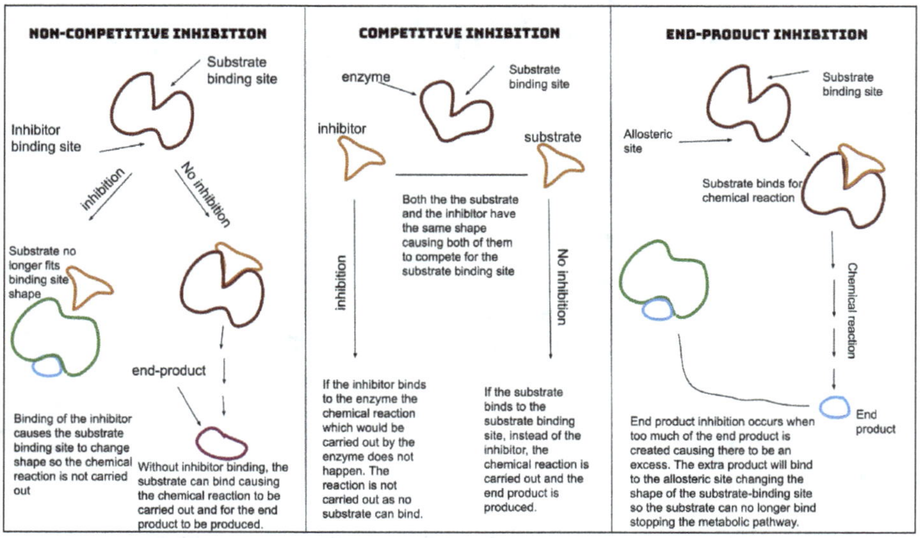

A figure comparing the three types of enzyme inhibitors and how they work in regards to substrate binding sites and inhibitors binding sites.

An enzyme binding site that would normally bind substrate can alternatively bind a competitive inhibitor, preventing substrate access. Dihydrofolate reductase is inhibited by methotrexate which prevents binding of its substrate, folic acid. Binding site in blue, inhibitor in green, and substrate in black.

An enzyme inhibitor is a molecule that binds to an enzyme and decreases its activity. Since blocking an enzyme's activity can kill a pathogen or correct a metabolic imbalance, many drugs are enzyme inhibitors. They are also used in pesticides. Not all molecules that bind to enzymes are inhibitors; *enzyme activators* bind to enzymes and increase their enzymatic activity, while enzyme substrates bind and are converted to products in the normal catalytic cycle of the enzyme.

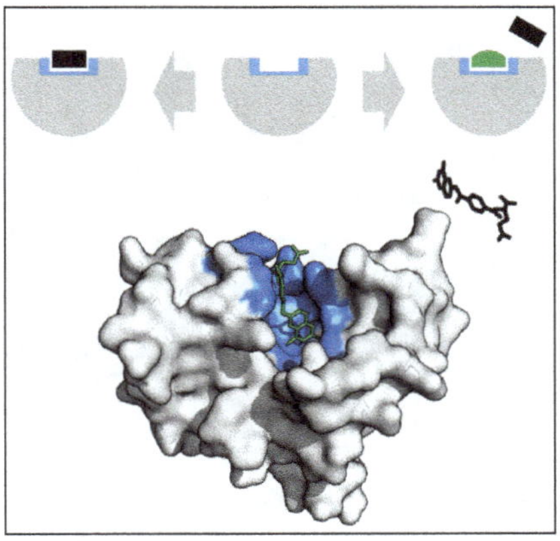

The binding of an inhibitor can stop a substrate from entering the enzyme's active site and hinder the enzyme from catalyzing its reaction. Inhibitor binding is either reversible or irreversible. Irreversible inhibitors usually react with the enzyme and change it chemically (e.g. via covalent bond formation). These inhibitors modify key amino acid residues needed for enzymatic activity. In contrast, reversible inhibitors bind non-covalently and different types of inhibition are produced depending on whether these inhibitors bind to the enzyme, the enzyme-substrate complex, or both.

Many drug molecules are enzyme inhibitors, so their discovery and improvement is an active area of research in biochemistry and pharmacology. A medicinal enzyme inhibitor is often judged by its specificity (its lack of binding to other proteins) and its potency (its dissociation constant, which indicates the concentration needed to inhibit the enzyme). A high specificity and potency ensure that a drug will have few side effects and thus low toxicity.

Enzyme inhibitors also occur naturally and are involved in the regulation of metabolism. For example, enzymes in a metabolic pathway can be inhibited by downstream products. This type of negative feedback slows the production line when products begin to build up and is an important way to maintain homeostasis in a cell. Other cellular enzyme inhibitors are proteins that specifically bind to and inhibit an enzyme target. This can help control enzymes that may be damaging to a cell, like proteases or nucleases.

A well-characterised example of this is the ribonuclease inhibitor, which binds to ribonucleases in one of the tightest known protein–protein interactions. Natural enzyme inhibitors can also be poisons and are used as defenses against predators or as ways of killing prey.

Reversible Inhibitors

Types of Reversible Inhibitors

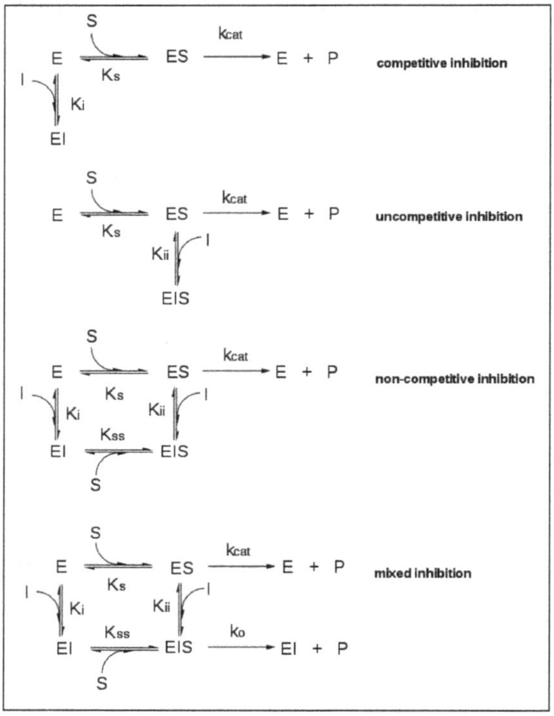

Types of inhibition. This classification was introduced by W.W. Cleland.

Reversible inhibitors attach to enzymes with non-covalent interactions such as hydrogen bonds, hydrophobic interactions and ionic bonds. Multiple weak bonds between the inhibitor and the active site combine to produce strong and specific binding. In contrast to substrates and irreversible inhibitors, reversible inhibitors generally do not undergo chemical reactions when bound to the enzyme and can be easily removed by dilution or dialysis.

There are four kinds of reversible enzyme inhibitors. They are classified according to the effect of varying the concentration of the enzyme's substrate on the inhibitor.

- In competitive inhibition, the substrate and inhibitor cannot bind to the enzyme at the same time, as shown in the figure on the right. This usually results from the inhibitor having an affinity for the active site of an enzyme where the substrate also binds; the substrate and inhibitor *compete* for access to the enzyme's active site. This type of inhibition can be overcome by sufficiently high concentrations of substrate (V_{max} remains constant), i.e., by out-competing the inhibitor. However, the apparent K_m will increase as it takes a higher concentration of the substrate to reach the K_m point, or half the V_{max}. Competitive inhibitors are often similar in structure to the real substrate.

- In uncompetitive inhibition, the inhibitor binds only to the substrate-enzyme complex. This type of inhibition causes V_{max} to decrease (maximum velocity decreases as a result of removing activated complex) and K_m to decrease (due to better binding efficiency as a result of Le Chatelier's principle and the effective elimination of the ES complex thus decreasing the K_m which indicates a higher binding affinity).
- In non-competitive inhibition, the binding of the inhibitor to the enzyme reduces its activity but does not affect the binding of substrate. As a result, the extent of inhibition depends only on the concentration of the inhibitor. V_{max} will decrease due to the inability for the reaction to proceed as efficiently, but K_m will remain the same as the actual binding of the substrate, by definition, will still function properly.
- In mixed inhibition, the inhibitor can bind to the enzyme at the same time as the enzyme's substrate. However, the binding of the inhibitor affects the binding of the substrate, and vice versa. This type of inhibition can be reduced, but not overcome by increasing concentrations of substrate. Although it is possible for mixed-type inhibitors to bind in the active site, this type of inhibition generally results from an allosteric effect where the inhibitor binds to a different site on an enzyme. Inhibitor binding to this allosteric site changes the conformation (i.e., tertiary structure or three-dimensional shape) of the enzyme so that the affinity of the substrate for the active site is reduced.

Quantitative Description of Reversible Inhibition

Reversible inhibition can be described quantitatively in terms of the inhibitor's binding to the enzyme and to the enzyme-substrate complex, and its effects on the kinetic constants of the enzyme. In the classic Michaelis-Menten scheme below, an enzyme (E) binds to its substrate (S) to form the enzyme–substrate complex ES. Upon catalysis, this complex breaks down to release product P and free enzyme. The inhibitor (I) can bind to either E or ES with the dissociation constants K_i or K_i', respectively.

- Competitive inhibitors can bind to E, but not to ES. Competitive inhibition increases K_m (i.e., the inhibitor interferes with substrate binding), but does not affect V_{max} (the inhibitor does not hamper catalysis in ES because it cannot bind to ES).
- Uncompetitive inhibitors bind to ES. Uncompetitive inhibition decreases both K_m' and 'V_{max}. The inhibitor affects substrate binding by increasing the enzyme's affinity for the substrate (decreasing K_m) as well as hampering catalysis (decreases V_{max}).

- Non-competitive inhibitors have identical affinities for E and ES (K_i = K_i'). Non-competitive inhibition does not change K_m (i.e., it does not affect substrate binding) but decreases V_{max} (i.e., inhibitor binding hampers catalysis).
- Mixed-type inhibitors bind to both E and ES, but their affinities for these two forms of the enzyme are different ($K_i \neq K_i$'). Thus, mixed-type inhibitors interfere with substrate binding (increase K_m) and hamper catalysis in the ES complex (decrease V_{max}).

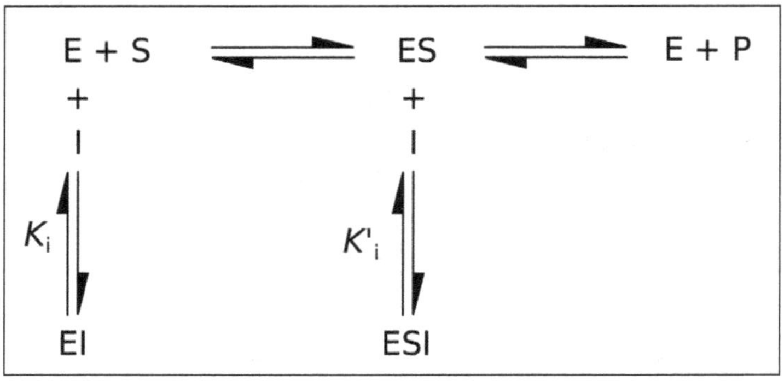

Kinetic scheme for reversible enzyme inhibitors.

When an enzyme has multiple substrates, inhibitors can show different types of inhibition depending on which substrate is considered. This results from the active site containing two different binding sites within the active site, one for each substrate. For example, an inhibitor might compete with substrate A for the first binding site, but be a non-competitive inhibitor with respect to substrate B in the second binding site.

Measuring the Dissociation Constants of a Reversible Inhibitor

An enzyme inhibitor is characterised by its two dissociation constants, K_i and K_i', to the enzyme and to the enzyme-substrate complex, respectively. The enzyme-inhibitor constant K_i can be measured directly by various methods; one extremely accurate method is isothermal titration calorimetry, in which the inhibitor is titrated into a solution of enzyme and the heat released or absorbed is measured. However, the other dissociation constant K_i' is difficult to measure directly, since the enzyme-substrate complex is short-lived and undergoing a chemical reaction to form the product. Hence, K_i' is usually measured indirectly, by observing the enzyme activity under various substrate and inhibitor concentrations, and fitting the data to a modified Michaelis–Menten equation,

$$V = \frac{V_{max}[S]}{\alpha K_m + \alpha'[S]} = \frac{(1/\alpha')V_{max}[S]}{(\alpha/\alpha')K_m + [S]}$$

where the modifying factors α and α' are defined by the inhibitor concentration and its two dissociation constants,

$$1+\frac{[I]}{K_i}$$

$$\alpha' = 1+\frac{[I]}{K_i'}.$$

Thus, in the presence of the inhibitor, the enzyme's effective K_m and V_{max} become (α/α') K_m and (1/α')V_{max}, respectively. However, the modified Michaelis-Menten equation assumes that binding of the inhibitor to the enzyme has reached equilibrium, which may be a very slow process for inhibitors with sub-nanomolar dissociation constants. In these cases, it is usually more practical to treat the tight-binding inhibitor as an irreversible inhibitor; however, it can still be possible to estimate K_i' kinetically if K_i is measured independently.

The effects of different types of reversible enzyme inhibitors on enzymatic activity can be visualized using graphical representations of the Michaelis–Menten equation, such as Lineweaver–Burk and Eadie-Hofstee plots. For example, in the Lineweaver–Burk plots at the right, the competitive inhibition lines intersect on the *y*-axis, illustrating that such inhibitors do not affect V_{max}. Similarly, the non-competitive inhibition lines intersect on the *x*-axis, showing these inhibitors do not affect K_m. However, it can be difficult to estimate K_i and K_i' accurately from such plots, so it is advisable to estimate these constants using more reliable nonlinear regression methods.

Special Cases

- The mechanism of partially competitive inhibition is similar to that of non-competitive, except that the EIS complex has catalytic activity, which may be lower or even higher (partially competitive activation) than that of the enzyme–substrate (ES) complex. This inhibition typically displays a lower V_{max}, but an unaffected K_m value.
- Uncompetitive inhibition occurs when the inhibitor binds only to the enzyme–substrate complex, not to the free enzyme; the EIS complex is catalytically inactive. This mode of inhibition is rare and causes a decrease in both V_{max} and the K_m value.
- Substrate and product inhibition is where either the substrate or product of an enzyme reaction inhibit the enzyme's activity. This inhibition may follow the competitive, uncompetitive or mixed patterns. In substrate inhibition there is a progressive decrease in activity at high substrate concentrations. This may indicate the existence of two substrate-binding sites in the enzyme. At low substrate, the high-affinity site is occupied and normal kinetics are followed. However, at higher concentrations, the second inhibitory site becomes occupied, inhibiting the enzyme. Product inhibition is often a regulatory feature in metabolism and can be a form of negative feedback.

- Slow-tight inhibition occurs when the initial enzyme–inhibitor complex EI undergoes isomerisation to a second more tightly held complex, EI*, but the overall inhibition process is reversible. This manifests itself as slowly increasing enzyme inhibition. Under these conditions, traditional Michaelis–Menten kinetics give a false value for K_i, which is time–dependent. The true value of K_i can be obtained through more complex analysis of the on (k_{on}) and off (k_{off}) rate constants for inhibitor association.

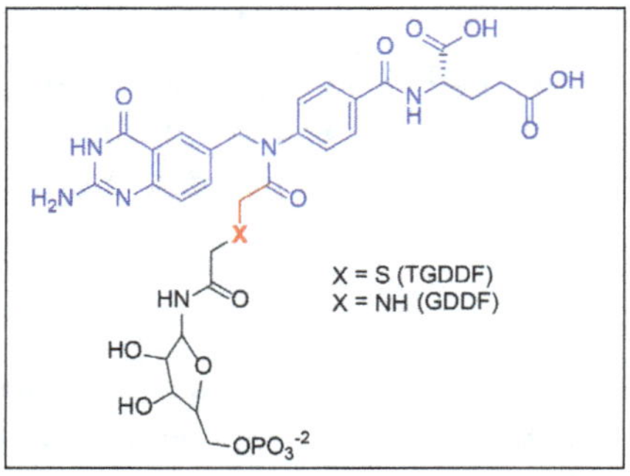

TGDDF/GDDF MAIs where blue depicts the tetrahydrofolate cofactor analog, black GAR or thioGAR and red, the connecting atoms.

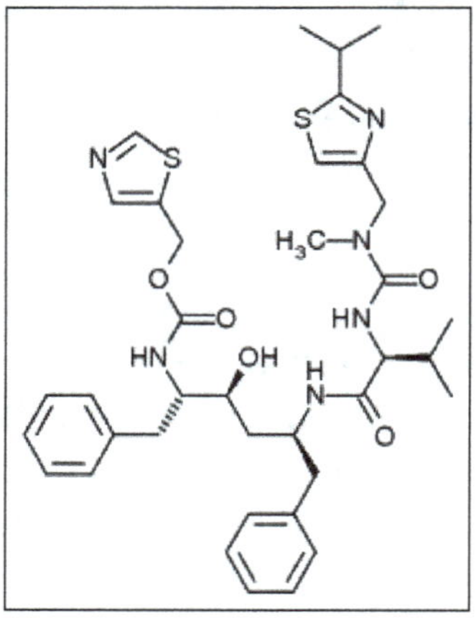

Peptide-based HIV-1 protease inhibitor ritonavir.

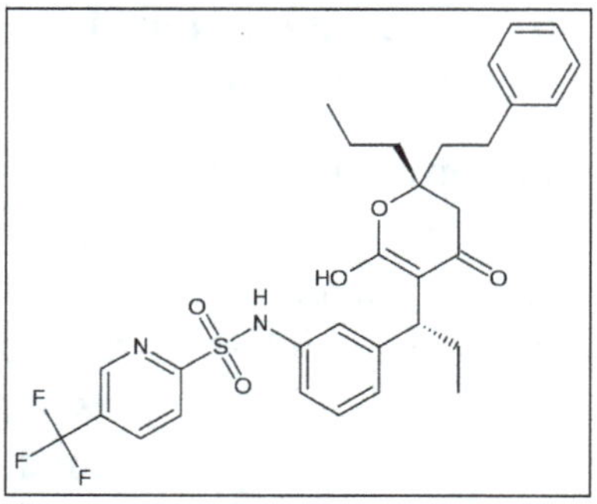

Nonpeptidic HIV-1 protease inhibitor tipranavir.

- Bi-substrate analog inhibitors are high affinity and selectivity inhibitors that can be prepared for enzymes that catalyze bi-molecular reactions by capturing the binding energy of each substrate into one molecule. For example, in the formyl transfer reactions of purine biosynthesis, a potent multi-substrate adduct inhibitor (MAI) to GAR TFase was prepared synthetically by linking analogs of the glycinamide ribonucleotide (GAR) substrate and the N-10-formyl tetrahydrofolate cofactor together to produce thioglycinamide ribonucleotide dideazafolate (TGDDF), or enzymatically from the natural GAR substrate to yield GDDF. Here the subnanomolar dissociation constant (KD) of TGDDF was greater than predicted presumably due to entropic advantages gained and positive interactions acquired through the atoms linking the components. MAIs have also been observed to be produced in cells by reactions of prodrugs such as isoniazid or enzyme inhibitor ligands (e.g., PTC124) with cellular cofactors such as NADH and ATP respectively.

Examples of Reversible Inhibitors

As enzymes have evolved to bind their substrates tightly, and most reversible inhibitors bind in the active site of enzymes, it is unsurprising that some of these inhibitors are strikingly similar in structure to the substrates of their targets. Inhibitors of DHFR are prominent examples. Other example of these substrate mimics are the protease inhibitors, a very successful class of antiretroviral drugs used to treat HIV. The structure of ritonavir, a protease inhibitor based on a peptide and containing three peptide bonds, is shown on the right. As this drug resembles the protein that is the substrate of the HIV protease, it competes with this substrate in the enzyme's active site.

Enzyme inhibitors are often designed to mimic the transition state or intermediate of an enzyme-catalyzed reaction. This ensures that the inhibitor exploits the transition state stabilising effect of the enzyme, resulting in a better binding affinity (lower K_i) than substrate-based designs. An example of such a transition state inhibitor is the antiviral drug oseltamivir; this drug mimics the planar nature of the ring oxonium ion in the reaction of the viral enzyme neuraminidase.

However, not all inhibitors are based on the structures of substrates. For example, the structure of another HIV protease inhibitor tipranavir is shown on the left. This molecule is not based on a peptide and has no obvious structural similarity to a protein substrate. These non-peptide inhibitors can be more stable than inhibitors containing peptide bonds, because they will not be substrates for peptidases and are less likely to be degraded.

In drug design it is important to consider the concentrations of substrates to which the target enzymes are exposed. For example, some protein kinase inhibitors have chemical structures that are similar to adenosine triphosphate, one of the substrates of these enzymes. However, drugs that are simple competitive inhibitors will have to compete with the high concentrations of ATP in the cell. Protein kinases can also be inhibited by competition at the binding sites where the kinases interact with their substrate proteins, and most proteins are present inside cells at concentrations much lower than the concentration of ATP. As a consequence, if two protein kinase inhibitors both bind in the active site with similar affinity, but only one has to compete with ATP, then the competitive inhibitor at the protein-binding site will inhibit the enzyme more effectively.

Irreversible Inhibitors

Types of Irreversible Inhibition (Covalent Inactivation)

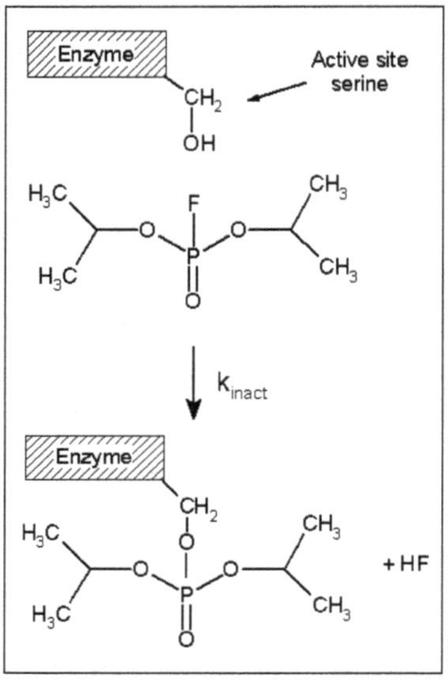

Reaction of the irreversible inhibitor diisopropylfluorophosphate (DFP) with a serine protease.

Irreversible inhibitors usually covalently modify an enzyme, and inhibition can therefore not be reversed. Irreversible inhibitors often contain reactive functional groups such as nitrogen mustards, aldehydes, haloalkanes, alkenes, Michael acceptors, phenyl sulfonates, or fluorophosphonates. These electrophilic groups react with amino acid side chains to form covalent adducts. The residues modified are those with side chains containing nucleophiles such as hydroxyl or sulfhydryl groups; these include the amino acids serine (as in DFP, right), cysteine, threonine, or tyrosine.

Irreversible inhibition is different from irreversible enzyme inactivation. Irreversible inhibitors are generally specific for one class of enzyme and do not inactivate all proteins; they do not function by destroying protein structure but by specifically altering the active site of their target. For example, extremes of pH or temperature usually cause denaturation of all protein structure, but this is a non-specific effect. Similarly, some non-specific chemical treatments destroy protein structure: for example, heating in concentrated hydrochloric acid will hydrolyse the peptide bonds holding proteins together, releasing free amino acids.

Irreversible inhibitors display time-dependent inhibition and their potency therefore cannot be characterised by an IC_{50} value. This is because the amount of active enzyme at a given concentration of irreversible inhibitor will be different depending on how long the inhibitor is pre-incubated with the enzyme. Instead, $k_{obs}/[I]$ values are used, where k_{obs} is the observed pseudo-first order rate of inactivation (obtained by plotting the log of % activity vs. time) and $[I]$ is the concentration of inhibitor. The $k_{obs}/[I]$ parameter is valid as long as the inhibitor does not saturate binding with the enzyme (in which case $k_{obs} = k_{inact}$).

Analysis of Irreversible Inhibition

As shown in the figure to the right, irreversible inhibitors have a short instance where they form a reversible non-covalent complex with the enzyme (EI or ESI) and this then reacts to produce the covalently modified "dead-end complex" EI* (an irreversible covalent complex). The rate at which EI* is formed is called the inactivation rate or k_{inact}. Since formation of EI may compete with ES, binding of irreversible inhibitors can be prevented by competition either with substrate or with a second, reversible inhibitor. This protection effect is good evidence of a specific reaction of the irreversible inhibitor with the active site.

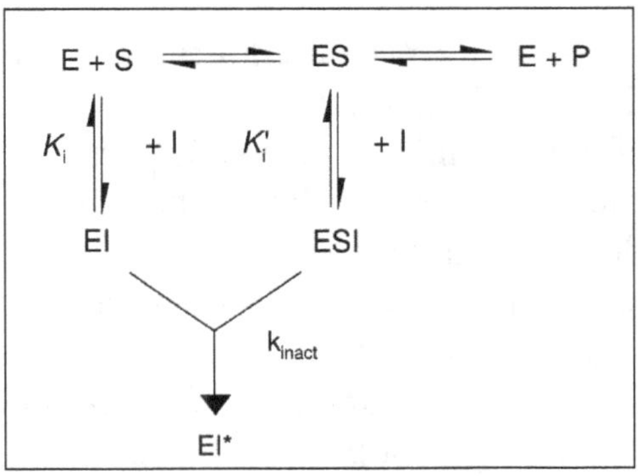

Kinetic scheme for irreversible inhibitors.

The binding and inactivation steps of this reaction are investigated by incubating the enzyme with inhibitor and assaying the amount of activity remaining over time. The activity will be decreased in a time-dependent manner, usually following exponential decay. Fitting these data to a rate equation gives the rate of inactivation at this concentration of inhibitor. This is done at several different concentrations of inhibitor. If a reversible EI complex is involved the inactivation rate will be saturable and fitting this curve will give k_{inact} and K_i.

Another method that is widely used in these analyses is mass spectrometry. Here, accurate measurement of the mass of the unmodified native enzyme and the inactivated enzyme gives the increase in mass caused by reaction with the inhibitor and shows the stoichiometry of the reaction. This is usually done using a MALDI-TOF mass spectrometer. In a complementary technique, peptide mass fingerprinting involves digestion of the native and modified protein with a protease such as trypsin. This will produce a set of peptides that can be analysed using a mass spectrometer. The peptide that changes in mass after reaction with the inhibitor will be the one that contains the site of modification.

Special Cases

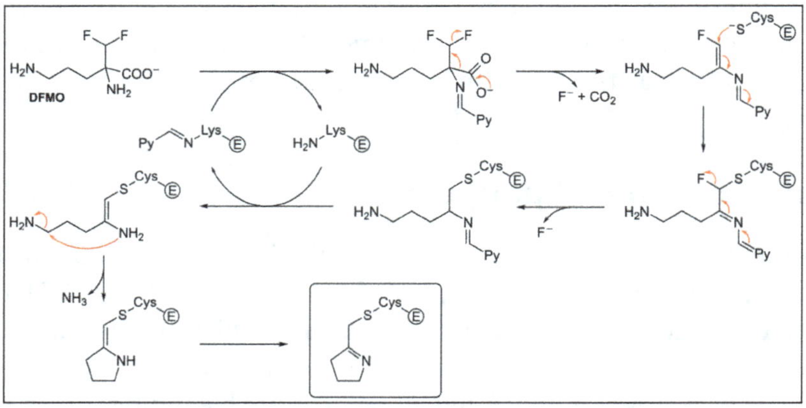

Chemical mechanism for irreversible inhibition of ornithine decarboxylase by DFMO. Pyridoxal 5'-phosphate (Py) and enzyme (E) are not shown.

Not all irreversible inhibitors form covalent adducts with their enzyme targets. Some reversible inhibitors bind so tightly to their target enzyme that they are essentially irreversible. These tight-binding inhibitors may show kinetics similar to covalent irreversible inhibitors. In these cases, some of these inhibitors rapidly bind to the enzyme in a low-affinity EI complex and this then undergoes a slower rearrangement to a very tightly bound EI* complex. This kinetic behaviour is called slow-binding. This slow rearrangement after binding often involves a conformational change as the enzyme "clamps down" around the inhibitor molecule. Examples of slow-binding inhibitors include some important drugs, such methotrexate, allopurinol, and the activated form of acyclovir.

Examples of Irreversible Inhibitors

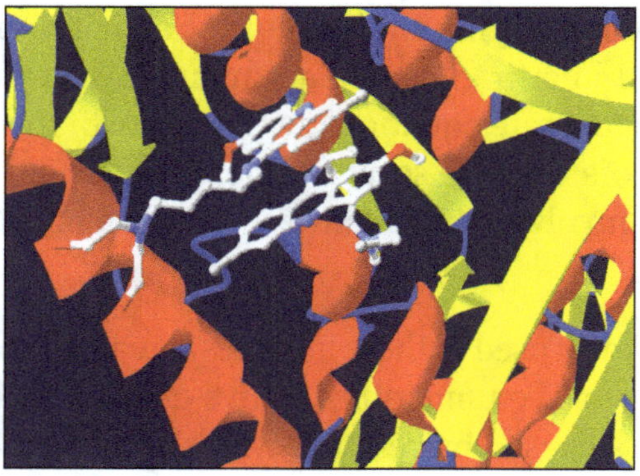

Trypanothione reductase with the lower molecule of an inhibitor bound irreversibly and the upper one reversibly.

Diisopropylfluorophosphate (DFP) is shown as an example of an irreversible protease inhibitor in the figure. The enzyme hydrolyses the phosphorus–fluorine bond, but the phosphate residue remains bound to the serine in the active site, deactivating it. Similarly, DFP also reacts with the active site of acetylcholine esterase in the synapses of neurons, and consequently is a potent neurotoxin, with a lethal dose of less than 100 mg.

Suicide inhibition is an unusual type of irreversible inhibition where the enzyme converts the inhibitor into a reactive form in its active site. An example is the inhibitor of polyamine biosynthesis, α-difluoromethylornithine or DFMO, which is an analogue of the amino acid ornithine, and is used to treat African trypanosomiasis (sleeping sickness). Ornithine decarboxylase can catalyse the decarboxylation of DFMO instead of ornithine, as shown above. However, this decarboxylation reaction is followed by the elimination of a fluorine atom, which converts this catalytic intermediate into a conjugated imine, a highly electrophilic species. This reactive form of DFMO then reacts with either a cysteine or lysine residue in the active site to irreversibly inactivate the enzyme.

Since irreversible inhibition often involves the initial formation of a non-covalent EI complex, it is sometimes possible for an inhibitor to bind to an enzyme in more than one way. For example, in the figure showing trypanothione reductase from the human protozoan parasite *Trypanosoma cruzi*, two molecules of an inhibitor called *quinacrine mustard* are bound in its active site. The top molecule is bound reversibly, but the lower one is bound covalently as it has reacted with an amino acid residue through its nitrogen mustard group.

Chemical Equilibrium

Chemical equilibrium is a condition in the course of a reversible chemical reaction in which no net change in the amounts of reactants and products occurs. A reversible chemical reaction is one in which the products, as soon as they are formed, react to produce the original reactants. At equilibrium, the two opposing reactions go on at equal rates, or velocities, hence there is no net change in the amounts of substances involved. At this point the reaction may be considered to be completed; i.e., for some specified reaction condition, the maximum conversion of reactants to products has been attained.

The conditions that pertain to equilibrium may be given quantitative formulation. For example, for the reversible reaction $A \rightleftharpoons B + C$, the velocity of the reaction to the right, r_1, is given by the mathematical expression (based on the law of mass action) $r_1 = k_1(A)$, where k_1 is the reaction-rate constant and the symbol in parentheses represents the concentration of A. The velocity of the reaction to the left, r_2, is $r_2 = k_2(B)(C)$. At equilibrium, $r_1 = r_2$, therefore:

$$k_1(A)_e = k_2(B)_e(C)_e \quad \text{or}$$
$$k_1 / k_2 = (B)_e(C)_e / (A)_e = k.$$

The subscript *e* represents conditions at equilibrium. For a given reaction, at some specified condition of temperature and pressure, the ratio of the amounts of products and reactants present at equilibrium, each raised to their respective powers, is a constant, designated the equilibrium constant of the reaction and represented by the symbol *K*. The value of the equilibrium constant varies with the temperature and pressure according to the principle of Le Chatelier.

By methods of statistical mechanics and chemical thermodynamics, it can be shown that the equilibrium constant is related to the change in the thermodynamic quantity called the standard Gibbs free energy accompanying the reaction. The standard Gibbs free energy of the reaction, $\Delta G°$, which is the difference between the sum of the standard free energies of the products and that of the reactants, is equal to the negative natural logarithm of the equilibrium constant multiplied by the so-called gas constant *R* and the absolute temperature *T*:

$$\Delta G^{\circ} = -RT\ in\ K.$$

The equation allows the calculation of the equilibrium constant, or the relative amounts of products and reactants present at equilibrium, from measured or derived values of standard free energies of substances.

References

- Schlögl, Robert (2015-03-09). "Heterogeneous Catalysis". Angewandte Chemie International Edition. 54 (11): 3465–3520. Doi:10.1002/anie.201410738. PMID 25693734
- Chemical-kinetics, science: britannica.com, Retrieved 31 March, 2020
- Espenson, James H. Chemical Kinetics and Reaction Mechanisms (2nd ed., mcgraw-Hill, 2002) chap.6, Deduction of Reaction Mechanisms ISBN 0-07-288362-6
- Investigation of chemical reactions in solution using API-MS Leonardo Silva Santos, Larissa Knaack, Jurgen O. Metzger Int. J. Mass Spectrom.; 2005; 246 pp 84 - 104; (Review) doi:10.1016/j.ijms.2005.08.016
- Iglesias, Pablo A.; Ingalls, Brian P. (2010). Control theory and systems biology. MIT Press. ISBN 9780262013345
- Bauer, Th.; Lunkenheimer, P.; Loidl, A. (2013). "Cooperativity and the Freezing of Molecular Motion at the Glass Transition". Physical Review Letters. 111 (22): 225702. Arxiv:1306.4630. Bibcode:2013phrvl.111v5702b. Doi:10.1103/physrevlett.111.225702. PMID 24329455
- Thomas, J. M.; Thomas, W. J. (2014-11-19). Principles and practice of heterogeneous catalysis (Second, revised ed.). Weinheim, Germany. ISBN 9783527683789. OCLC 898421752
- Chemical-equilibrium, science: britannica.com, Retrieved 14 July, 2020

Electrochemistry 3

- **Electrochemical Reaction**
- **Electrochemical Potential**
- **Electrochemical Cell**
- **Electrolysis**
- **Applications of Electrochemistry**

Electrochemistry is the branch of physical chemistry which is concerned with the study of electricity and its interaction with chemical reactions. Galvanic cell, Daniell cell, primary and secondary cell are a few types of electrochemical cells used in electroplating. The topics elaborated in this chapter will help in gaining a better perspective about electrochemistry.

Electrochemistry is the study of chemical processes that cause electrons to move. This movement of electrons is called electricity, which can be generated by movements of electrons from one element to another in a reaction known as an oxidation-reduction ("redox") reaction. A redox reaction is a reaction that involves a change in oxidation state of one or more elements.

When a substance loses an electron, its oxidation state increases; thus, it is oxidized. When a substance gains an electron, its oxidation state decreases, thus being reduced. For example, for the redox reaction

$$H_2 + F_2 \rightarrow 2\,HF$$

can be rewritten as follows:

- Oxidation reaction

$$H_2 \rightarrow 2\,H^+ + 2e^-$$

- Reduction reaction:

$$F_2 + 2e^- \rightarrow 2\,F^-$$

- Overall Reaction

  $$H_2+F_2 \rightarrow 2H^{+}+2F^{-}$$

Oxidation is the loss of electrons, whereas reduction refers to the acquisition of electrons, as illustrated in the respective reactions above. The species being oxidized is also known as the reducing agent or reductant, and the species being reduced is called the oxidizing agent or oxidant. In this case, H_2 is being oxidized (and is the reducing agent), while F_2 is being reduced (and is the oxidizing agent).

Example: Iron-vanadium couple.

Given the redox reaction $Fe^{3+}+V^{2+} \rightarrow Fe^{2+}+V^{3+}$, which species is oxidized? Which is reduced? Identify the reducing agent and the oxidizing agent.

Solution:

Fe^{3+} is *reduced* into Fe^{2+} and V^{2+} is *oxidized* into V^{3+}. This is because the oxidized species loses electrons, and the reduced species gains electrons.

Iron gains an electron $\left(Fe^{3+} \rightarrow Fe^{2+}\right)$, and vanadium loses an electron $\left(V^{2+} \rightarrow V^{3+}\right)$.

Thus, Fe3+Fe3+ is the oxidizing agent and V^{2+} is the reducing agent.

Oxidation States

Rules for assigning oxidation states:

- Free elements have an oxidation state of 0. (e.g., He, N_2, O_2 has an oxidation state of 0).
- The oxidation state of one atom ion must equal the net charge. (e.g., F^- oxidation state is -1, K^+ oxidation state is +1).
- The sum of the oxidation state has to equal the total net charge for a compound. (e.g., MnO_4 has a net charge of -1, Mn(+7)O_4(-8)= -1).
- The alkali metals (Group I elements) have an oxidation state of +1. (e.g., Li_2O, Li= +1).
- The alkaline earth metals (Group II elements) always have an oxidation state of +2. (e.g., CaO, Ca=+2).
- Oxygen has an oxidation state of -2 in a compound.
- Fluorine has an oxidation state of -1 in a compound.
- Hydrogen has an oxidation state of +1 in a compound.
- Transition metals and other metals may have more than one common ionic charge. (e.g., Chromium's common ionic charges are Cr^{+2} and Cr^{+3}).

Balancing Redox Reactions

Oxidation Number Method

- Step 1: Assign oxidation numbers to each atom.
- Step 2: Determine the net change in charge to determine the ratio of atoms.
- Step 3: Use the ratio to eliminate the net charge change.
- Step 4: Use the ratio as coefficients for the elements.
- Step 5: Add H^+ (under acidic conditions), OH^- (under basic conditions), and H_2O to balance charges.

Half-reaction Method

- Step 1: Determine oxidation numbers for each atom.
- Step 2: Use oxidation numbers to determine what is oxidized and what is reduced.
- Step 3: Write a half-reaction for reduction.
- Step 4: Write a half-reaction for oxidation.
- Step 5: Balance all elements except H and O.
    - If have acid redox reaction: Balance the O using H_2O, balance the H using protons.
    - If have base redox reaction: Balance O using OH^-.
- Step 6: Add up the charge on each side.
- Step 7: Balance the charges by adding electrons.
- Step 8: Multiply the half-reactions by factors that cancel out electrons.
- Step 9: Add the two half-reactions back together to eliminate out intermediates.

Example: Manganese.

Balance the following reaction in an acidic aqueous solution,

$$MnO_4^- + H_2C_2O_4 \rightarrow Mn^2 + + CO_2$$

Solution:

- Reduction half-reaction:

  $2 \times (5e^- + 8H+ + MnO_4^- \rightarrow Mn^{2+} + 4H_2O)$

- Oxidation half-reaction:

$$5\times(H_2C_2O_4 \rightarrow CO_2 + 2H^+ + 2e^-)$$

- Combining and canceling gives the following:

$$\cancel{10e^-} + 16H^+ + 2MnO_4^{1-} + 5H_2C_2O_4 \rightarrow 2Mn^{2+} + 8H_2O + 5CO_2 + 10H^+ + \cancel{10e^-}$$

Answer:

$$6H^+ + 2MnO_4^{1-} + 5H_2C_2O_4 \rightarrow 2Mn^{2+} + 8H_2O + 5CO_2$$

Electrochemical Reaction

Electrochemical reaction is any process either caused or accompanied by the passage of an electric current and involving in most cases the transfer of electrons between two substances—one a solid and the other a liquid.

Under ordinary conditions, the occurrence of a chemical reaction is accompanied by the liberation or absorption of heat and not of any other form of energy; but there are many chemical reactions that—when allowed to proceed in contact with two electronic conductors, separated by conducting wires—liberate what is called electrical energy, and an electric current is generated. Conversely, the energy of an electric current can be used to bring about many chemical reactions that do not occur spontaneously. A process involving the direct conversion of chemical energy when suitably organized constitutes an electrical cell. A process whereby electrical energy is converted directly into chemical energy is one of electrolysis; i.e., an electrolytic process. By virtue of their combined chemical energy, the products of an electrolytic process have a tendency to react spontaneously with one another, reproducing the substances that were reactants and were therefore consumed during the electrolysis. If this reverse reaction is allowed to occur under proper conditions, a large proportion of the electrical energy used in the electrolysis may be regenerated. This possibility is made use of in accumulators or storage cells, sets of which are known as storage batteries. The charging of an accumulator is a process of electrolysis; a chemical change is produced by the electric current passing through it. In the discharge of the cell, the reverse chemical change occurs, the accumulator acting as a cell that produces an electric current.

General Principles

Substances that are reasonably good conductors of electricity may be divided into two groups: the metallic, or electronic, conductors and the electrolytic conductors. The metals and many nonmetallic substances such as graphite, manganese dioxide, and lead sulfide exhibit metallic conductivity; the passage of an electric current through them produces heating and magnetic effects but no chemical changes. Electrolytic conductors, or electrolytes, comprise most acids, bases, and salts, either in the molten condition or in solution in water or other solvents. Plates or rods composed of a suitable metallic conductor dipping into the fluid electrolyte are employed to conduct the current into and out of the liquid; i.e., to act as electrodes. When a current is passed between electrodes through an electrolyte, not only are heating and magnetic effects produced but also definite chemical changes occur. At or in the neighbourhood of the negative electrode, called the cathode, the chemical change may be the deposition of a metal or the liberation of hydrogen and formation of a basic substance or some other chemical reduction process; at the positive electrode, or anode, it may be the dissolution of the anode itself, the liberation of a nonmetal, the production of oxygen and an acidic substance, or some other chemical oxidation process.

An electrolyte, prepared either by the melting of a suitable substance or by the dissolving of it in water or other liquid, owes its characteristic properties to the presence in it of electrically charged atoms or groups of atoms produced by the spontaneous splitting up or dissociation of the molecules of the substance. In solutions of the so-called strong electrolytes, most of the original substance, or in some solutions perhaps all of it, has undergone this process of electrolytic dissociation into charged particles, or ions. When an electrical potential difference (i.e., a difference in degree of electrification) is established between electrodes dipping into an electrolyte, positively charged ions move toward the cathode and ions bearing negative charges move toward the anode. The electric current is carried through the electrolyte by this migration of the ions. When an ion reaches the electrode of opposite polarity, its electrical charge is donated to the metal, or an electric charge is received from the metal. The ion is thereby converted into an ordinary neutral atom or group of atoms. It is this discharge of ions that gives rise to one of the types of chemical changes occurring at electrodes.

The Electrochemical Process

Interactions of matter associated with the passage of an electric current depend upon the characteristics of the negatively charged electron. As the basic particle of electricity, the electron has an affinity for positively charged particles of matter, protons, whether in atoms, groups of atoms, or molecules. This affinity is analogous to the chemical affinity that particles exhibit among themselves. In fact, all chemical reactions result from a shift in the electron structure of atoms, and free electrons can combine with particles of matter (reduction) or be released by them (oxidation). The quantitative relationship between the free electrons of an electric current and the particles of a substance in which they cause a reaction is defined by the laws of Faraday. Electrons are available in large quantities in a relatively free (mobile) state only in substances called electronic conductors, among which metals are the most important. Thus, an electron conductor must be present as a basic component of any system in which electrochemical reactions are to occur. Furthermore, the availability of electrons in a conductor is limited by energy distribution to such an extent that electrochemical reactions take place only in the immediate vicinity of the electronic conductor's surface—i.e., a few angstroms from the conductor into the solution. These reactions are, therefore, normally considered as occurring at the interface, or common boundary, between an electronic conductor, such as an electrode, and an ionic conductor of electricity, such as an electrolytic solution. Electrochemical reaction will take place, however, only to the extent that electricity can flow through such a system as a whole. To achieve this, it is necessary for the system to form a closed loop, electronically speaking.

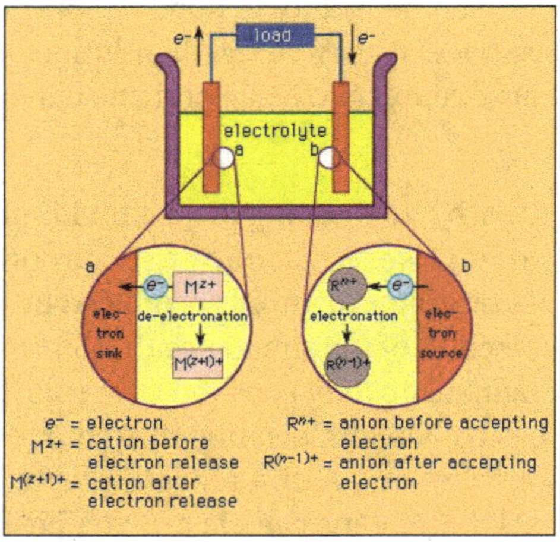

The mechanism of electron movement in an electrochemical cell.

To summarize, if at one metal-solution interface electrons are coming out of the metal, reducing a component of the solution, there must exist a second metal-solution interface where electrons are going into the metal in the process of oxidation.

The two electrodes and the ionic conductor in between (e.g., an aqueous solution of some ionized salt) represent an electrochemical cell. The process occurring in the cell as a whole is a redox process with the reduction of one species spatially separated from the oxidation of another one. As a consequence of Faraday's law, the rates of electrochemical reactions at electrodes (expressed in terms of gram moles per second per square centimetre of electrode surface) are directly proportional to the current density (expressed in amperes per square centimetre)—i.e., current flowing through the cell divided by the electrode surface area.

Sites of Electrochemical Reactions

Electrochemical reactions take place where the electron conductor meets the ionic conductor—i.e., at the electrode–electrolyte interface. Characteristic of this region, considered to be a surface phase, is the existence of a specific structure of particles and the presence of an electric field of considerable intensity (up to 10,000,000 volts per centimetre) across it; the field is caused by the separation of charges that are present between the two bulk phases in contact. For most purposes the surface phase can be considered as a parallel plate condenser, with one plate on the centre of the ions that have been brought to the electrode, at the distance of their closest approach to it, and with the second plate at the metal surface; between the two plates and acting as a dielectric (i.e., a nonconducting material) are oriented water molecules. This structure is termed the electric double layer and is illustrated in figure.

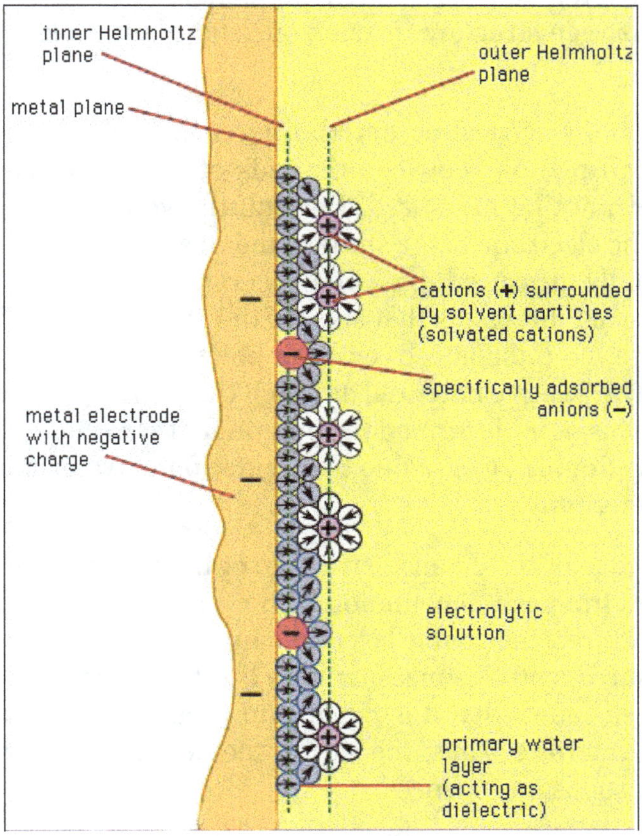

Double-layer structure and change of potential with distance from the electrode surface.

Thermal motion of the positive ions in the solution makes the condenser plate on the electrolyte side of the interface diffuse—i.e., the ions are distributed in a cloudlike way. This condition justifies the division of the potential change between the bulk of metal and the bulk of electrolyte into two parts: first, that between the metal surface and the first ionic layer at the distance of closest approach (called the outer Helmholtz plane, in which the ions are usually surrounded by solvent particles—i.e., are solvated); and second, that between the first ionic layer and the bulk of the solution, the diffuse part of the double layer. The picture is further complicated by the presence of ions in the electrode surface layer in addition to those that are present for electrostatic reasons—i.e., by the force of attraction or repulsion between electric charges. Such electrode surface layer ions are said to be specifically adsorbed on the electrode surface. Since this species of ions is attracted by the surface to a distance closer than the "distance of the closest approach" of ions, further subdivision of the inner part of the electric double layer is justified. Hence, the inner Helmholtz plane is introduced as the plane formed by the centres of specifically adsorbed ions. Adsorption of neutral molecules on the surface can also change the properties of the electric double layer. This change occurs as a consequence of replacing the water molecules, and thus changes that part of the potential (electrical) difference across the double layer that is caused by the adsorbed dipoles (water molecules that have a polarity—i.e., they behave like minute magnets—because of their hydrogen-oxygen structure, making one end of the molecule positive and the other end negative).

The absolute value of electrical potential difference, symbolized in calculation by the letters delta and psi, $\Delta\psi$, between the bulk of a metal electrode and the bulk of an electrolyte cannot be measured. Instead, the voltage of a special cell, composed of the specific electrode being studied and of an arbitrarily selected reference electrode, is normally measured; the voltage is referred to as the relative electrode potential, E. Of special interest is that state of the electrode at which there is no net charge (in this case, no unbalanced, or extra positive, charge) at the metal side of the double layer. The relative potential at which this state is achieved is characteristic of each metal. This point is termed the potential of zero charge. At that potential, the field across the double layer is due to orientation of water molecules and other dipoles at the surface only.

Most of the knowledge of the detailed structure of the interface between a metal and an electrolyte arises from experimentation with mercury, the only metal that is liquid at ordinary temperatures; the double layer structure turns out to have surface tensions that must be measured, and this measurement is difficult with solid metals. By 1970, however, it had been shown that it is possible to measure surface tension changes at the metal-solution interface. Thus, the way to the determination of the double layer structure involving solids was opened.

Substances that are semiconductors can also be employed as electron carriers in electrochemical reactions. Semiconductors are substances which range between serving as insulators at low temperatures and as metallic-type conductors at high temperatures. In the case of semiconductors, however, the electric double layer has a more complex structure inasmuch as the condenser plate at the electrode side of the double layer also becomes diffuse. Thus, the overall potential difference between bulks of the phases in contact comprises also the potential difference between the bulk of the semiconductor and its surface.

Types of Electrochemical Reactions

There are several types of electrochemical reactions.

Simple Redox Reactions

A simple redox reaction is one that involves a change in the electrical charge of a charge carrier, usually a simple or complex ion in the solution, by its taking away, an electron from the electrode (reduction), or its giving an electron to the electrode (oxidation). The same carrier may be present in solution in two states of charge. The higher, more positive charge is called the oxidized state, and the lower, less positive charge is called the reduced state. For example, when ferric and ferrous ions are both present in solution in significant quantity, and when electron exchange with the electrode is sufficiently fast, redox equilibrium is established at the electrode, giving it a well-defined potential, or reversible redox potential.

Reactions that Produce Gases

When hydrogen ions in solution react with electrons ejected from a metal, hydrogen atoms are formed at the surface, where they combine among themselves or with other hydrogen ions and electrons to give gaseous hydrogen molecules. If all the reactions are fast enough, an equilibrium is attained between hydrogen ions and gaseous hydrogen. A metal in contact with solution at which such a situation exists is called the reversible hydrogen electrode, and its electrical potential is arbitrarily taken to be zero; every other electrode can thus be compared with it as it represents the basis for constituting the hydrogen scale of relative electrode potentials. Similarly, negative hydroxyl ions in solution (OH^-) can be made to give up electrons to a metal and, in a series of reactions, the final one is the formation of gaseous oxygen. Chlorine is another gaseous product; it evolves upon electrochemical oxidation of chloride ions in concentrated solutions of neutral and acid salts.

Reactions that Deposit and Dissolve Metals

When a metal ion is reduced and discharged as a neutral atom, or species, it tends to build into the metal lattice of the electrode. Thus, metals can be deposited at electrodes. Conversely, if electrons are taken away from the metal electrode by applying positive potentials to it, the metal ions thus formed can cross the double layer of electric charge at the interface, undergo hydration (combination with water), and enter the solution. The metal electrode thus dissolves. Many metals establish well-defined electric potentials when they are in contact with their own ions in solution.

Oxidation and Reduction of Organic Compounds

A reaction of the oxidation and reduction of organic compounds can also be done at electrodes. Such reactions, however, are mostly irreversible in the literal sense that they lead to products that cannot easily be converted back into the original substance. Exceptions are some oxygen- and nitrogen-containing compounds (quinones, amines, and nitrous compounds) that can give fairly well-defined reversible potentials.

Mechanism of Charge Transfer

The causes of the thermodynamically irreversible behaviour of electrode reactions are found in the nature of the elementary act of charge transfer. Like any chemical reaction, this act is inhibited by the existence of an energy barrier between the oxidized and the reduced state. This barrier implies that the reaction could take place only in the special circumstances when, during the course of numerous interactions with other species (atoms, ions, molecules, etc.) surrounding it, a molecule attains an excited state in which it has an abnormal energy content. In most chemical reactions, this energy content must be sufficient for the species to come into what is called the transition state; the transition state characterizes the top of the energy barrier just before a reaction begins. If such a model is applied to electron transfer at an interface, calculation shows that electron exchange reactions at electrodes would be prohibitively slow, a conclusion at variance with the observed phenomena; quantum mechanical laws, however, govern the motion of electrons, and their inclusion changes the calculations to fit reality. Quantum mechanics require that for fast electron exchange to take place, electrons in a particle outside the double layer (e.g., a hydrated ion at the outer Helmholtz plane) must attain certain well-defined quantized energy levels equal to those in which free electrons exist in the metal. Since such states can be attained by the particle at a lower energy-content than that needed for its transfer over the top of the energy barrier, according to the classical view, this fast process of electron exchange between the electrode and a particle in solution is termed electron tunnelling through the energy barrier.

Whereas the rate of chemical processes, or what may loosely be termed the speed of reaction, can be influenced only by changing the concentrations of reactants or by changing the temperature or both, the rate of electrochemical processes also can be manipulated by changing the electrode potential. Making the electrode more negative increases the number of electrons in the metal ready to tunnel to ions, and hence the rate of the reduction process increases. Conversely, making the potential more positive decreases this rate and increases the number of particles ready to give away electrons, thus increasing the rate of the oxidation process.

It can be deduced that there must exist a direct proportionality between the rate of reaction and the concentration of the reacting species and at the same time an exponential proportionality between the rate of reaction and the electrode potential.

At any electrode potential, both reduction of one species and oxidation of the product of reduction are taking place but at different rates; the rate of each reaction is determined by the respective concentration and by the corresponding effects of potential. The rate of an electrochemical reaction can best be described as the electric current density—i.e., a measure of the quantity of electrons moving in a certain volume of space during a specified unit of time.

The relationships can be represented quantitatively by an equation in which the net, or resulting, current (the difference of the rate of electron ejection across the interface to particles in solution, diminished by the rate at which particles in solution inject electrons into the metal) is equated to the difference of the rates of reduction and oxidation and the variables and constants that relate to these reactions.

Complex Electrochemical Reactions

Electrochemical processes considered so far involve simple reactions of a particle with a single electron to produce a reduced ion (e.g., the ferrous ion of iron with two positive charges, Fe^{++}), or vice versa. Such are the simple ionic redox processes, where the only difference in structure between a reactant and a reaction product may be due to some rearrangement of the neighbouring solvent molecules. When one or more transfers of electrons between the electrode and a species in solution are accompanied by major structural changes (e.g., when hydroxyl ions, OH^-, transform into a molecule of oxygen, O_2, and a water molecule, H_2O, in the process of oxygen evolution at the anode, or positive electrode), the reaction usually consists of a sequence of events, called elementary acts, or unit steps, constituting the reaction mechanism. Intermediate states between the steps usually involve some unstable intermediate species with higher energy content than those of the reactants or of the reaction products.

Complex reaction mechanisms can consist of a number of electron transfer steps, with some chemical steps preceding or succeeding the electron transfer steps or taking place in between them. Most organic electrochemical reactions are complex, involving large numbers of electrons in the overall reaction. Usually one step in the reaction encounters the largest energy barrier. The rate of occurrence of this step limits the rate of the overall reaction (i.e., all other steps must occur at the same net rate, although they could provide for a much faster overall change). This step is called the rate-determining step and, for most practical purposes, all intermediate steps before and after it can be considered to be in equilibrium. (It is interesting to note that whenever this is the case, the Butler-Volmer equation is applicable, but with specific values of the transfer coeffi-cients αa and αc characteristic of the mechanism of the reaction).

Types of Complex Reactions

Electrocatalysis

The problems related to the increase of rates of electrochemical reactions, or, to put it another way, the decrease of overpotential, needed to perform reactions at a given rate are the subject of electrocatalysis. Both increase and decrease are of considerable practical importance since they affect the economics of electrochemical processes. Electrocatalysis is concerned with the electrode as a substrate, or base, for electrochemical reaction and with the effect of its bulk and surface properties on the rate of reaction. Contrary to expectation, the rate of the basic step in electrochemical reaction—electron transfer—is independent of the ease with which electrons are released from the metal. Hence, those simple electrode processes in which no other changes but electron transfer take place have heats of activation virtually independent of the metal substrate.

In a reaction involving the formation of a chemical bond between the electrode substrate and one of the radicals (charged particles) formed on the surface, rates of reaction at a given potential may vary for different substrates by many orders of magnitude. The variation is a function of the strength (or energy) of bonds established between the intermediate species and the surface, and when the substrates are transition metals, regular relations between reaction rate and certain characteristics of the metal's electronic structure are observed. If such bonds are too weak, the catalytic effect of the substrate is small. If they are too strong, the intermediate species is too stable to react further, and the surface of the metal becomes virtually blocked for further reaction. Thus, there is an optimum bond strength that can be seen if the rate of a reaction is plotted against a property of the metal substrate, which is proportional to the ability of the surface to form bonds—e.g., its heat of sublimation.

Electrochemical reactions can also be catalyzed or inhibited by foreign species present in the electrolytic solution. In the most general case, ions can change the rates of reactions by changing the properties of the double layer by specific adsorption effects. Organic molecules inhibit electrochemical reaction by blocking the surface. For some organic ions or dipoles, the blocking effect can begin at a certain well-defined potential. Before attaining it, the electrode reaction gives the usual current density-potential relationship, which is then suddenly interrupted by an abnormal fall of the current, corresponding to adsorption.

Electrocrystallization

Deposition of metals and other substances at electrodes as a consequence of an electrode process exhibits a number of specific features. The electrode process is followed by crystal building, and this results in a continuous change of the electrode surface. This change, in turn, affects the electrochemical properties of the system—the double-layer capacity and the rate constants of the charge transfer processes. Hence, if electrochemical properties are to be studied, transient methods should be employed that allow measurements to be made before major changes in surface morphology (structure) take place.

Since the two steps, the discharge of ions at the electrode and the incorporation of the discharged ions into the crystal lattice, are separated in time and space, an intermediate species exists at the surface, that of relatively loosely bound and freely moving atoms, called adatoms. Since the electrons tend to join the rest in the bulk of the metal, adatoms appear to have a partial charge, less than that of the elementary positive charge. The adatoms therefore attract solvent molecules, and the species is partially solvated. This reaction justifies considering an adatom as a kind of adsorbed ion, called an adion, which, however, has already undergone partial discharge.

Because such an intermediate state is possible, ions need not be reduced to the neutral atomic state at the point of incorporation into the crystal lattice. Indeed, energy considerations reveal this reduction to be improbable. Instead, discharge of ions is favoured on crystal planes. Their motion from the location of discharge to a site on the crystal occurs by surface diffusion. When such surface diffusion is inhibited, it can, in some cases, control the rate of reaction.

Problems of the crystal-building step in electrocrystallization are in many respects identical with those of crystal growth from the gas phase. Crystal building can occur in any one of three possible ways.

Kinks, naturally existing at any metal surface, form a suitable half-lattice position at which an atom is surrounded by one-half of the number of atoms that would surround it in the bulk of the metal; there, adatoms can be successively trapped and thus the crystal lattice is extended along a crystal edge and further on across the surface. This step growth mechanism is shown in figure. The mechanism, however, has a limited capacity for crystal growth. A step can move as far as the edge of a crystal, and step growth would lead to smoothing of the surface to perfection, but then further growth would cease.

Mechanisms associated with screw dislocations, or twinning edges, can provide for a continuous growth of crystals. The screw dislocation mechanism, shown in figure, is made possible by a specific fault often found in the crystal lattice that may be called a dislocation originating from a shift of one atom in the lattice with respect to a perfect arrangement. This shift may then result in the formation of a monoatomic edge projecting above the electrode surface at which new atoms can be stacked. The stacking produces a turning of the edge around the base atom as a centre, the process producing a spiral growth.

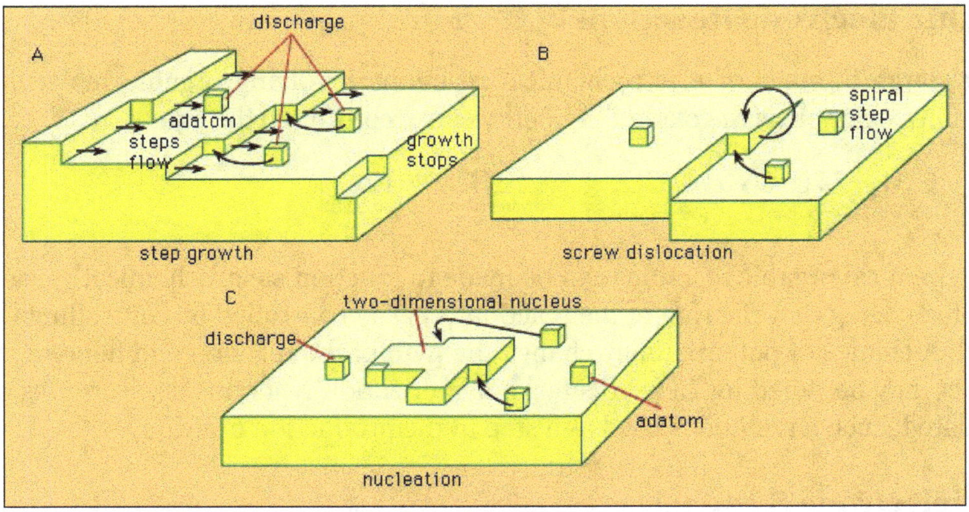

Mechanisms of crystal growth.

If growth sites are rare, or if the substrate at which deposition should take place is foreign to the depositing metal, the charge transfer results in an accumulation of adatoms to a concentration considerably larger than that which can exist there at equilibrium with the crystal lattice. In such a situation, termed supersaturation, agglomeration of adatoms to form crystal nuclei is favoured. Surface energy requirements show that, at any degree of supersaturation, nuclei of certain dimensions are stable and can represent sites for further growth, as shown in figure.

The formation of deposits is controlled by the above mechanism as long as the discharge process supplies ample amounts of adatoms as building material over the entire surface. If the rate of deposition is increased, so as to produce near the surface considerable depletion of ions, uneven deposition will start. This is caused by the protrusions of a normally rough metal surface being closer to the bulk of solution than the recessed parts and, hence, getting a somewhat faster supply of the discharging species. Once such a situation is established, it tends to develop further. The faster growing points penetrate into ever richer layers of solution, resulting in ever faster growth. Thus, a natural consequence of deposition under transport-controlled rate is the amplification of the original surface roughness and the appearance of protruding spikes, called dendrites, as deposits.

A converse process is the smoothing of the original surface irregularities, which may occur when some foreign species, called an additive, is present in solution and adsorbs on the surface and inhibits the process of discharge. If those molecules are incorporated into a growing deposit, a situation may arise in which their supply to recessed parts of the surface becomes slower than to elevated parts. As a result, deposition becomes faster at recessed parts than at elevated ones and leveling of the surface occurs. This process has considerable technical application. Metal dissolution can sometimes be governed by a similar, transport-controlled mechanism, in which case a polishing effect is obtained. Electropolishing has found wide application in practice.

Organic Electrode Reactions

A very large number of electrochemical reactions involving organic molecules are known. An example is the oxidation of ethylene according to the equation:

$$\underset{\text{(ethylene)}}{C_2H_4} + \underset{\text{(water)}}{4H_2O} \rightarrow \underset{\text{(carbon dioxide)}}{2CO_2} + \underset{\text{(hydrogen ion)}}{12H^+} + \underset{\text{(electrons)}}{12e^-} .$$

Many chemical organic reactions can be made to function electrochemically, a general advantage being that the rate of the reaction is easily controlled by controlling the potential. A change of potential may change the path of the reaction and hence a certain product may be tuned in. In particular, polymerization reactions at electrodes can be stimulated; such a method is used as a step in the production of nylon.

Multielectrode Systems

So far, systems have been considered in which a single electrode process takes place. In principle, at any electrode potential all species present in the system fall into two categories: those that are stable, and those that undergo oxidation or reduction. The stable species are those that at the given electrode potential would not decrease their free energy by giving off or accepting electrons.

At any potential there should occur a codischarge of all unstable species. Thus, the system can be considered as a multielectrode system, consisting of as many electrodes as there are redox couples present. The rate at which different processes occur, however, can be so widely different that usually a single process is by far the dominant one. Systems at which two electrode processes occur at comparable rates are of considerable importance. If two kinds of metal ions are discharged simultaneously, an alloy is formed upon crystallization. The properties of the alloy would in most cases be those determined by the phase diagram (a plot of the temperature of melting versus the composition of the mixed system) for the ratio of quantities of discharged metals given by the rates at which the discharges take place. In many cases, nonequilibrium metal phases are formed giving unusual properties to the alloy.

If a process of metal dissolution (an oxidation process) can occur at a rate comparable to that of some reduction process on the same metal, a corrosion couple is established. Thus, zinc immersed in acid solution tends to establish a potential sufficiently positive for the metal to eject metal ions into the solution (oxidation). At the same time, it is also sufficiently negative for the reduction of hydrogen ions, present in any aqueous solution, to hydrogen gas. Hence, a spontaneous process of hydrogen evolution and of dissolution of the metal will take place. The mixed potential, spontaneously established by the corroding metal, is obtained by equating expressions for the anodic and cathodic currents of the two processes in the corrosion couple.

Corrosion can be prevented in two ways. One is by using an external source to make the potential of the metal sufficiently negative to bring it into the potential region in which the metal is stable, called cathodic protection; and the other is by provoking by some means the formation of a film on the surface that would slow the process. Such films could consist of an oxide or a layer of organic molecules that prevents dissolution and hence is called an inhibitor.

Applications

Electrochemical processes are used in many ways and their use is likely to increase because they can replace polluting chemical situations with nonpolluting electrochemical ones. In many fields, however, applications have been profitable for some time. Major categories are listed below.

Metallurgy

All technologically important metals, except iron and steel, are either obtained or refined by electrochemical processes; for example, aluminum, titanium, alkaline earth, and alkali metals are obtained by electrodeposition from molten salts, and copper is refined by electrolysis in aqueous copper sulfate solutions.

Electroplating

One of the major ways of both decorating objects and improving their resistance to corrosion is by electroplating them. All major metal-working industries, particularly the automobile industry, have large electroplating plants.

Chemical Industry

Electrolysis of brine to obtain chlorine and caustic soda is an electrochemical process that has become one of the largest volume productions in the chemical industry. Modern processes cover a wide field, from the production of a variety of inorganic compounds to the production of such synthetic fibres as nylon. Intensive research in organic electrochemistry promises major developments in application, particularly with the prospect of greatly reduced electricity costs expected eventually to arise from the development of controlled fusion.

Batteries

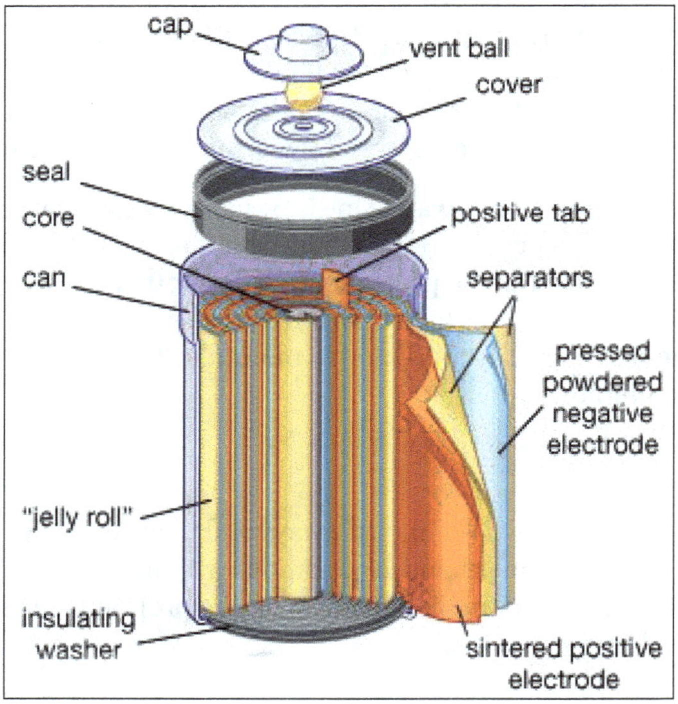

Nickel (hydroxide)-cadmium cell of "jelly roll" construction. This rechargable battery is commonly used in portable devices.

Electrochemical storage of electricity is effected in batteries. Such devices are electrochemical cells and consist of two electrodes per unit. As the electricity to be stored is accepted on the plates of the cell, it converts substances on the plates to new substances having a higher energy than the old ones. When it is desired to make the electricity available again, the terminals of the battery are connected to the load and the substances on the battery plates retransform themselves to those originally present, giving off electricity as a product of their electrochemical reactions. The steadily rising production of the lead-acid battery is largely the result of its use for starting the internal-combustion engine, which has had an equally steady rise. Other electrochemical systems are also used as storers. The nickel-iron (Edison cell) and nickel-cadmium battery with alkaline electrolyte are both used in applications where longer lives than those of the lead-acid battery are needed; the silver-zinc battery is used to start airplane engines because of its high power per unit of weight. A variety of new systems is being investigated for covering other needs. One of the greatest challenges to electrochemists and electrochemical engineers is that of producing a battery with sufficient power and energy density to run an automobile the way gasoline (petrol) does. Even if the best hypothetical predictions for removal of polluting chemicals from automobile exhausts is realized, the cleanup will not be sufficient because the expected growth of the automobile population will continue to increase the pollutant rate.

Fuel cells

The energy of chemical reactions is converted into electrical energy in fuel cells. In these, the fuel (e.g., hydrogen, hydrazine) is fed continuously to one electrode, while oxygen from the air is reacting at the other one. The efficiency of energy conversion in fuel cells is more than twice that attainable by conventional means—for example, by means of internal combustion.

Analytical chemistry

In analytical chemistry, most modern automated instrumental analysis is based on electrode processes—for example, potentiometry, used to measure ionization constant.

Electrochemical Potential

In electrochemistry, the electrochemical potential, μ, sometimes abbreviated to ECP, is a thermodynamic measure of chemical potential that does not omit the energy contribution of electrostatics. Electrochemical potential is expressed in the unit of J/mol.

Each chemical species (for example, "water molecules", "sodium ions", "electrons", etc.) has an electrochemical potential (a quantity with units of energy) at any given point in space, which represents how easy or difficult it is to add more of that species to that location. If possible, a species will move from areas with higher electrochemical potential to areas with lower electrochemical potential; in equilibrium, the electrochemical potential will be constant everywhere for each species (it may have a different value for different species). For example, if a glass of water has sodium ions (Na^+) dissolved uniformly in it, and an electric field is applied across the water, then the sodium ions will tend to get pulled by the electric field towards one side. We say the ions have electric potential energy, and are moving to lower their potential energy. Likewise, if a glass of water has a lot of dissolved sugar on one side and none on the other side, each sugar molecule will randomly diffuse around the water, until there is equal concentration of sugar everywhere. We say that the sugar molecules have a "chemical potential", which is higher in the high-concentration areas, and the molecules move to lower their chemical potential. These two examples show that an electrical potential and a chemical potential can both give the same result: A redistribution of the chemical species. Therefore, it makes sense to combine them into a single "potential", the electrochemical potential, which can directly give the net redistribution taking both into account.

It is (in principle) easy to measure whether or not two regions (for example, two glasses of water) have the same electrochemical potential for a certain chemical species (for example, a solute molecule): Allow the species to freely move back and forth between the two regions (for example, connect them with a semi-permeable membrane that lets only that species through). If the chemical potential is the same in the two regions, the species will occasionally move back and forth between the two regions, but on average there is just as much movement in one direction as the other, and there is zero net migration (this is called "diffusive equilibrium"). If the chemical potentials of the two regions are different, more molecules will move to the lower chemical potential than the other direction.

Moreover, when there is not diffusive equilibrium, i.e., when there is a tendency for molecules to diffuse from one region to another, then there is a certain free energy released by each net-diffusing molecule. This energy, which can sometimes be harnessed (a simple example is a concentration cell), and the free-energy per mole is exactly equal to the electrochemical potential difference between the two regions.

In generic terms, electrochemical potential is the mechanical work done in bringing 1 mole of an ion from a standard state to a specified concentration and electrical potential. According to the IUPAC definition, it is the partial molar Gibbs energy of the substance at the specified electric potential, where the substance is in a specified phase. Electrochemical potential can be expressed as

$$\bar{\mu}_i = \mu_i + z_i F\Phi$$

where:

- μ_i is the electrochemical potential of species i, in J/mol,
- μ_i is the chemical potential of the species i, in J/mol,
- z_i is the valency (charge) of the ion i, a dimensionless integer,
- F is the Faraday constant, in C/mol,

Φ is the local electrostatic potential, in V.

In the special case of an uncharged atom, $z_i = 0$, and so $\mu_i = \mu_i$.

Electrochemical potential is important in biological processes that involve molecular diffusion across membranes, in electroanalytical chemistry, and industrial applications such as batteries and fuel cells. It represents one of the many interchangeable forms of potential energy through which energy may be conserved.

In cell membranes, the electrochemical potential is the sum of the chemical potential and the membrane potential.

Electrochemical Cell

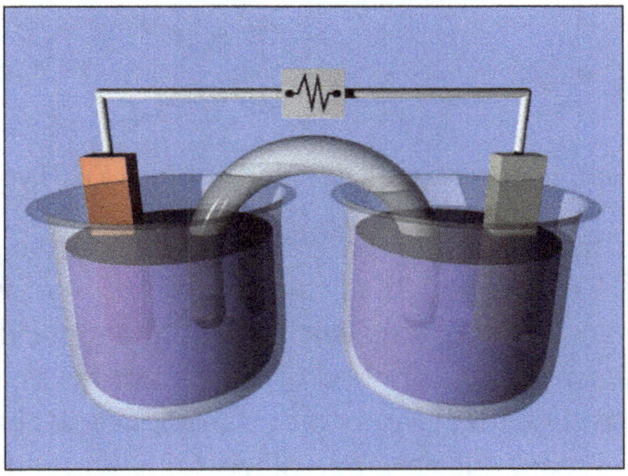

A demonstration electrochemical cell setup resembling the Daniell cell. The two half-cells are linked by a salt bridge carrying ions between them. Electrons flow in the external circuit.

An electrochemical cell is a device capable of either generating electrical energy from chemical reactions or using electrical energy to cause chemical reactions. The electrochemical cells which generate an electric current are called voltaic cells or galvanic cells and those that generate chemical reactions, via electrolysis for example, are called electrolytic cells. A common example of a galvanic cell is a standard 1.5 volt cell meant for consumer use. A battery consists of one or more cells, connected either in parallel, series or series-and-parallel pattern.

Electrolytic Cell

An electrolytic cell is an electrochemical cell that drives a non-spontaneous redox reaction through the application of electrical energy. They are often used to decompose chemical compounds, in a process called electrolysis.

Important examples of electrolysis are the decomposition of water into hydrogen and oxygen, and bauxite into aluminium and other chemicals. Electroplating (e.g. of copper, silver, nickel or chromium) is done using an electrolytic cell. Electrolysis is a technique that uses a direct electric current (DC).

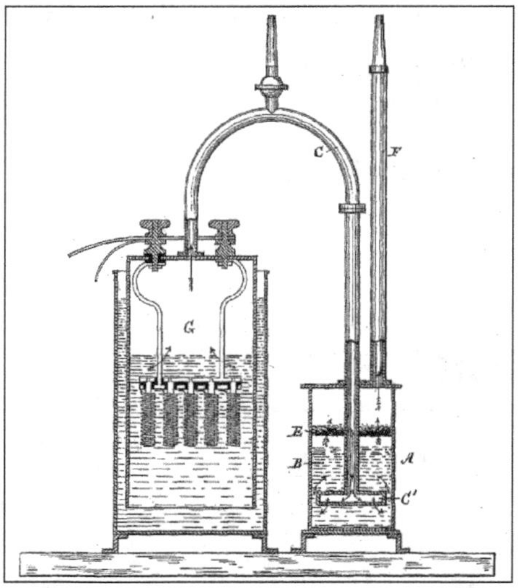

Nineteenth century electrolytic cell for producing oxyhydrogen.

An electrolytic cell has three component parts: an electrolyte and two electrodes (a cathode and an anode). The electrolyte is usually a solution of water or other solvents in which ions are dissolved. Molten salts such as sodium chloride are also electrolytes. When driven by an external voltage applied to the electrodes, the ions in the electrolyte are attracted to an electrode with the opposite charge, where charge-transferring (also called faradaic or redox) reactions can take place. Only with an external electrical potential (i.e. voltage) of correct polarity and sufficient magnitude can an electrolytic cell decompose a normally stable, or inert chemical compound in the solution. The electrical energy provided can produce a chemical reaction which would not occur spontaneously otherwise.

Fuel cell

A fuel cell is an electrochemical cell that converts the chemical energy from a fuel into electricity through an electrochemical reaction of hydrogen fuel with oxygen or another oxidizing agent. Fuel cells are different from batteries in requiring a continuous source of fuel and oxygen (usually from air) to sustain the chemical reaction, whereas in a battery the chemical energy comes from chemicals already present in the battery. Fuel cells can produce electricity continuously for as long as fuel and oxygen are supplied.

The first fuel cells were invented in 1838. The first commercial use of fuel cells came more than a century later in NASA space programmes to generate power for satellites and space capsules. Since then, fuel cells have been used in many other applications. Fuel cells are used for primary and backup power for commercial, industrial and residential buildings and in remote or inaccessible areas. They are also used to power fuel cell vehicles, including forklifts, automobiles, buses, boats, motorcycles and submarines.

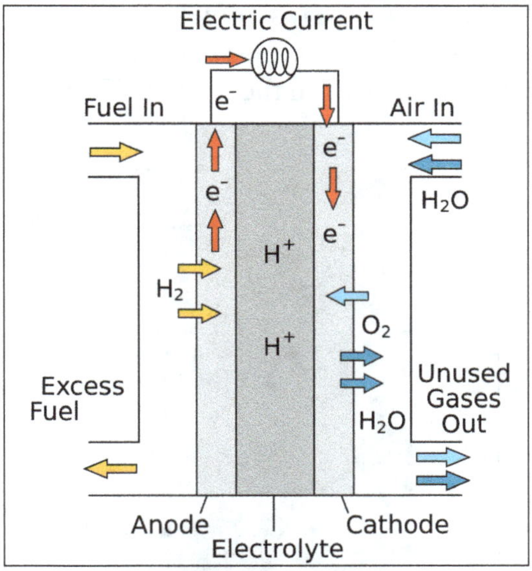

Scheme of a proton-conducting fuel cell.

There are many types of fuel cells, but they all consist of an anode, a cathode, and an electrolyte that allows positively charged hydrogen ions (protons) to move between the two sides of the fuel cell. At the anode a catalyst causes the fuel to undergo oxidation reactions that generate protons (positively charged hydrogen ions) and electrons. The protons flow from the anode to the cathode through the electrolyte after the reaction. At the same time, electrons are drawn from the anode to the cathode through an external circuit, producing direct current electricity. At the cathode, another catalyst causes hydrogen ions, electrons, and oxygen to react, forming water. Fuel cells are classified by the type of electrolyte they use and by the difference in startup time ranging from 1 second for proton exchange membrane fuel cells (PEM fuel cells, or PEMFC) to 10 minutes for solid oxide fuel cells (SOFC). A related technology is flow batteries, in which the fuel can be regenerated by recharging. Individual fuel cells produce relatively small electrical potentials, about 0.7 volts, so cells are "stacked", or placed in series, to create sufficient voltage to meet an application's requirements. In addition to electricity, fuel cells produce water, heat and, depending on the fuel source, very small amounts of nitrogen dioxide and other emissions. The energy efficiency of a fuel cell is generally between 40–60%; however, if waste heat is captured in a cogeneration scheme, efficiencies up to 85% can be obtained.

The fuel cell market is growing, and in 2013 Pike Research estimated that the stationary fuel cell market will reach 50 GW by 2020.

Half-cells

An electrochemical cell consists of two half-cells. Each half-cell consists of an electrode and an electrolyte. The two half-cells may use the same electrolyte, or they may use different electrolytes. The chemical reactions in the cell may involve the electrolyte, the electrodes, or an external substance (as in fuel cells that may use hydrogen gas as a reactant). In a full electrochemical cell, species from one half-cell lose electrons (oxidation) to their electrode while species from the other half-cell gain electrons (reduction) from their electrode.

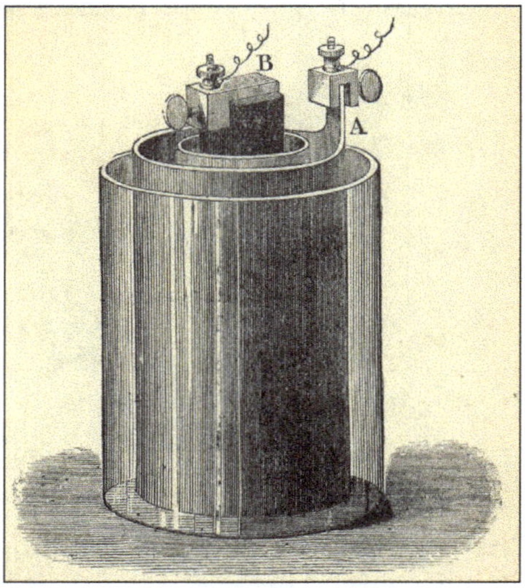

The Bunsen cell, invented by Robert Bunsen.

A salt bridge (e.g., filter paper soaked in KNO_3, NaCl, or some other electrolyte) is often employed to provide ionic contact between two half-cells with different electrolytes, yet prevent the solutions from mixing and causing unwanted side reactions. An alternative to a salt bridge is to allow direct contact (and mixing) between the two half-cells, for example in simple electrolysis of water.

As electrons flow from one half-cell to the other through an external circuit, a difference in charge is established. If no ionic contact were provided, this charge difference would quickly prevent the further flow of electrons. A salt bridge allows the flow of negative or positive ions to maintain a steady-state charge distribution between the oxidation and reduction vessels, while keeping the contents otherwise separate. Other devices for achieving separation of solutions are porous pots and gelled solutions. A porous pot is used in the Bunsen cell (right).

Equilibrium Reaction

Each half-cell has a characteristic voltage. Various choices of substances for each half-cell give different potential differences. Each reaction is undergoing an equilibrium reaction between different oxidation states of the ions: When equilibrium is reached, the cell cannot provide further voltage. In the half-cell that is undergoing oxidation, the closer the equilibrium lies to the ion/atom with the more positive oxidation state the more potential this reaction will provide. Likewise, in the reduction reaction, the closer the equilibrium lies to the ion/atom with the more negative oxidation state the higher the potential.

Cell Potential

The cell potential can be predicted through the use of electrode potentials (the voltages of each half-cell). These half-cell potentials are defined relative to the assignment of 0 volts to the standard hydrogen electrode (SHE).. The difference in voltage between electrode potentials gives a prediction for the potential measured. When calculating the difference in voltage, one must first rewrite the half-cell reaction equations to obtain a balanced oxidation-reduction equation.

- Reverse the reduction reaction with the smallest potential (to create an oxidation reaction/overall positive cell potential).
- Half-reactions must be multiplied by integers to achieve electron balance.

Cell potentials have a possible range of roughly zero to 6 volts. Cells using water-based electrolytes are usually limited to cell potentials less than about 2.5 volts due to high reactivity of the powerful oxidizing and reducing agents with water that is needed to produce a higher voltage. Higher cell potentials are possible with cells using other solvents instead of water. For instance, lithium cells with a voltage of 3 volts are commonly available.

The cell potential depends on the concentration of the reactants, as well as their type. As the cell is discharged, the concentration of the reactants decreases and the cell potential also decreases.

Galvanic Cell

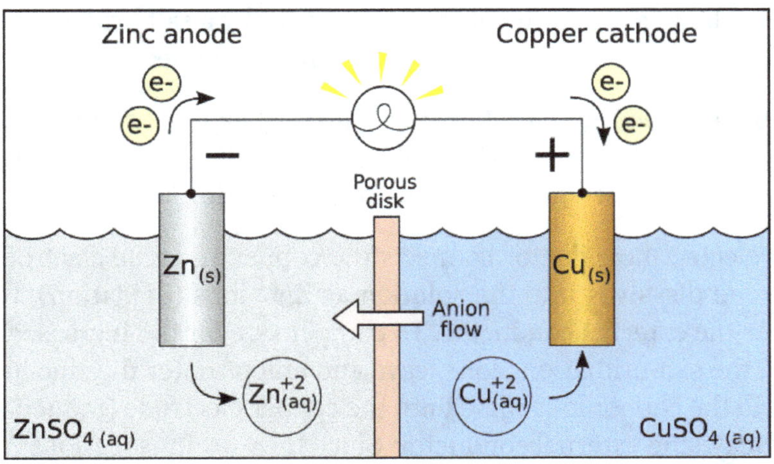

Galvanic cell with no cation flow.

A galvanic cell or voltaic cell, named after Luigi Galvani or Alessandro Volta, respectively, is an electrochemical cell that derives electrical energy from spontaneous redox reactions taking place within the cell. It generally consists of two different metals immersed in an electrolyte, or of individual half-cells with different metals and their ions in solution connected by a salt bridge or separated by a porous membrane.

Volta was the inventor of the voltaic pile, the first electrical battery. In common usage, the word "battery" has come to include a single galvanic cell, but a battery properly consists of multiple cells.

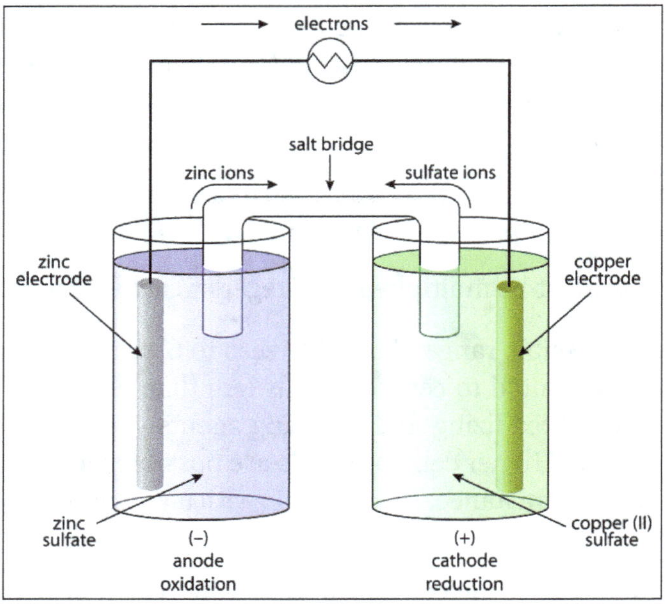

Schematic of Zn-Cu galvanic cell.

In its simplest form, a half-cell consists of a solid metal (called an electrode) that is submerged in a solution; the solution contains cations (+) of the electrode metal and anions (–) to balance the charge of the cations. The full cell consists of two half-cells, usually connected by a semi-permeable membrane or by a salt bridge that prevents the ions of the more noble metal from plating out at the other electrode.

A specific example is the Daniell cell, with a zinc (Zn) half-cell containing a solution of $ZnSO_4$ (zinc sulfate) and a copper (Cu) half-cell containing a solution of $CuSO_4$ (copper sulfate). A salt bridge is used here to complete the electric circuit.

If an external electrical conductor connects the copper and zinc electrodes, zinc from the zinc electrode dissolves into the solution as Zn^{2+} ions (oxidation), releasing electrons that enter the external conductor. To compensate for the increased zinc ion concentration, via the salt bridge zinc ions leave and anions enter the zinc half-cell. In the copper half-cell, the copper ions plate onto the copper electrode (reduction), taking up electrons that leave the external conductor. Since the Cu^{2+} ions (cations) plate onto the copper electrode, the latter is called the cathode. Correspondingly the zinc electrode is the anode. The electrochemical reaction is:

$$Zn + Cu^{2+} \rightarrow Zn^{2+} + Cu$$

In addition, electrons flow through the external conductor, which is the primary application of the galvanic cell.

The emf of the cell is the difference of the half-cell potentials, a measure of the relative ease of dissolution of the two electrodes into the electrolyte. The emf depends on both the electrodes and on the electrolyte, an indication that the emf is chemical in nature.

Electrochemical Thermodynamics of Galvanic Cell Reactions

The electrochemical processes in a galvanic cell occur because reactants of high free energy (e.g. metallic Zn and hydrated Cu^{2+} in the Daniell cell) are converted to lower-energy products (metallic Cu and hydrated Zn^{2+} in this example). The difference in the lattice cohesive energies of the electrode metals is sometimes the dominant energetic driver of the reaction, specifically in the Daniell cell. Metallic Zn, Cd, Li, and Na, which are not stabilized by d-orbital bonding, have higher cohesive energies (i.e. they are more weakly bonded) than all transition metals, including Cu, and are therefore useful as high-energy anode metals.

The difference between the metals' ionization energies in water is the other energetic contribution that can drive the reaction in a galvanic cell; it is not important in the Daniell cell because the energies of hydrated Cu^{2+} and Zn^{2+} ions happen to be similar. Both atom transfer, e.g. of zinc from the metal electrode into the solution, and electron transfer, from metal atoms or to metal ions, play important roles in a galvanic cell. Concentration cells, whose electrodes and ions are made of the same metal and which are driven by an entropy increase and free-energy decrease as ion concentrations equalize, show that the electronegativity difference of the metals is not the driving force of electrochemical processes.

Galvanic cells and batteries are typically used as a source of electrical power. The energy derives from a high-cohesive-energy metal dissolving while to a lower-energy metal is deposited, and/or from high-energy metal ions plating out while lower-energy ions go into solution.

Quantitatively, the electrical energy produced by a galvanic cell is approximately equal to the standard free-energy difference of the reactants and products, denoted as $\Delta_r G^o$. In the Daniell cell, most of the electrical energy of $\Delta_r G^o = -213$ kJ/mol can be attributed to the -207 kJ/mol difference between Zn and Cu lattice cohesive energies.

Half Reactions and Conventions

A half-cell contains a metal in two oxidation states. Inside an isolated half-cell, there is an oxidation-reduction (redox) reaction that is in chemical equilibrium, a condition written symbolically as follows (here, "M" represents a metal cation, an atom that has a charge imbalance due to the loss of "n" electrons):

$$M^{n+}\left(\text{oxidized species}\right) + ne^- \rightleftharpoons M\,(\text{reduced species})$$

A galvanic cell consists of two half-cells, such that the electrode of one half-cell is composed of metal A, and the electrode of the other half-cell is composed of metal B; the redox reactions for the two separate half-cells are thus:

$$A^{n+} + ne^- \rightleftharpoons A$$

$$B^{m+} + me^- \rightleftharpoons B$$

The overall balanced reaction is:

$$mA + nB^{m+} \rightleftharpoons nB + mA^{n+}$$

In other words, the metal atoms of one half-cell are oxidized while the metal cations of the other half-cell are reduced. By separating the metals in two half-cells, their reaction can be controlled in a way that forces transfer of electrons through the external circuit where they can do useful work.

- The electrodes are connected with a metal wire in order to conduct the electrons that participate in the reaction.

  In one half-cell, dissolved metal-B cations combine with the free electrons that are available at the interface between the solution and the metal-B electrode; these cations are thereby neutralized, causing them to precipitate from solution as deposits on the metal-B electrode, a process known as plating.

  This reduction reaction causes the free electrons throughout the metal-B electrode, the wire, and the metal-A electrode to be pulled into the metal-B electrode. Consequently, electrons are wrestled away from some of the atoms of the metal-A electrode, as though the metal-B cations were reacting directly with them; those metal-A atoms become cations that dissolve into the surrounding solution.

  As this reaction continues, the half-cell with the metal-A electrode develops a positively charged solution (because the metal-A cations dissolve into it), while the other half-cell develops a negatively charged solution (because the metal-B cations precipitate out of it, leaving behind the anions); unabated, this imbalance in charge would stop the reaction. The solutions of the half-cells are connected by a salt bridge or a porous plate that allows ions to pass from one solution to the other, which balances the charges of the solutions and allows the reaction to continue.

By definition:

- The anode is the electrode where oxidation (loss of electrons) takes place (metal-A electrode); in a galvanic cell, it is the negative electrode, because when oxidation occurs, electrons are left behind on the electrode. These electrons then flow through the external circuit to the cathode (positive electrode) (while in electrolysis, an electric current drives electron flow in the opposite direction and the anode is the positive electrode).

- The cathode is the electrode where reduction (gain of electrons) takes place (metal-B electrode); in a galvanic cell, it is the positive electrode, as ions get reduced by taking up electrons from the electrode and plate out (while in electrolysis, the cathode is the negative terminal and attracts positive ions from the solution). In both cases, the statement 'the cathode attracts cations' is true.

Galvanic cells, by their nature, produce direct current. The Weston cell has an anode composed of cadmium mercury amalgam, and a cathode composed of pure mercury. The electrolyte is a (saturated) solution of cadmium sulfate. The depolarizer is a paste of mercurous sulfate. When the electrolyte solution is saturated, the voltage of the cell is very reproducible; hence, in 1911, it was adopted as an international standard for voltage.

A battery is a set of galvanic cells that are connected together to form a single source of voltage. For instance, a typical 12V lead–acid battery has six galvanic cells connected in series with the anodes composed of lead and cathodes composed of lead dioxide, both immersed in sulfuric acid. Large battery rooms, for instance in a telephone exchange providing central office power to user's telephones, may have cells connected in both series and parallel.

Cell Voltage

The voltage (electromotive force E°) produced by a galvanic cell can be estimated from the standard Gibbs free energy change in the electrochemical reaction according to

$$E^o_{\text{cell}} = -\Delta_r G^o / (\nu_e F)$$

where ν_e is the number of electrons transferred in the balanced half reactions, and F is Faraday's constant. However, it can be determined more conveniently by the use of a standard potential table for the two half cells involved. The first step is to identify the two metals and their ions reacting in the cell. Then one looks up the standard electrode potential, E°, in volts, for each of the two half reactions. The standard potential of the cell is equal to the more positive E° value minus the more negative E° value.

For example, in the figure above the solutions are $CuSO_4$ and $ZnSO_4$. Each solution has a corresponding metal strip in it, and a salt bridge or porous disk connecting the two solutions and allowing SO_4^{2-} ions to flow freely between the copper and zinc solutions. To calculate the standard potential one looks up copper and zinc's half reactions and finds;

$$Cu^{2+} + 2e^- \rightleftharpoons Cu \qquad E^o = +0.34\text{ V}$$

$$Zn^{2+} + 2e^- \rightleftharpoons Zn \qquad E^o = -0.76\text{ V}$$

Thus the overall reaction is:

$$Cu^{2+} + Zn \rightleftharpoons Cu + Zn^{2+}$$

The standard potential for the reaction is then +0.34 V – (−0.76 V) = 1.10 V. The polarity of the cell is determined as follows. Zinc metal is more strongly reducing than copper metal because the standard (reduction) potential for zinc is more negative than that of copper. Thus, zinc metal will lose electrons to copper ions and develop a positive electrical charge. The equilibrium constant, K, for the cell is given by:

$$\mathrm{n}\,K = \frac{\nu_e F E^o_{\text{cell}}}{RT}$$

where F is the Faraday constant, R is the gas constant and T is the temperature in kelvins. For the Daniell cell K is approximately equal to 1.5×10^{37}. Thus, at equilibrium, a few electrons are transferred, enough to cause the electrodes to be charged.

Actual half-cell potentials must be calculated by using the Nernst equation as the solutes are unlikely to be in their standard states,

$$E_{\text{half-cell}} = E^o - \frac{RT}{\nu_e F}\ln_e Q$$

where Q is the reaction quotient. When the charges of the ions in the reaction are equal, this simplifies to,

$$E_{\text{half-cell}} = E^o - 2.303\frac{RT}{\nu_e F}\log_{10}\left\{M^{n+}\right\}$$

where $\left\{M^{n+}\right\}$ is the activity of the metal ion in solution. In practice concentration in mol/L is used in place of activity. The metal electrode is in its standard state so by definition has unit activity. The potential of the whole cell is obtained as the difference between the potentials for the two half-cells, so it depends on the concentrations of both dissolved metal ions. If the concentrations are the same, $E_{\text{cell}} = E^o_{\text{cell}}$ and the Nernst equation is not needed under the conditions assumed here.

The value of 2.303 $\frac{R}{F}$ is 1.9845×10^{-4} V/K, so at 25 °C (298.15 K) the half-cell potential will change by only 0.05918 V/ν_e if the concentration of a metal ion is increased or decreased by a factor of 10.

$$E_{\text{half-cell}} = E^o - \frac{0.05918\text{ V}}{\nu_e}\log_{10}\left[M^{n+}\right]$$

These calculations are based on the assumption that all chemical reactions are in equilibrium. When a current flows in the circuit, equilibrium conditions are not achieved and the cell voltage will usually be reduced by various mechanisms, such as the development of overpotentials. Also, since chemical reactions occur when the cell is producing power, the electrolyte concentrations change and the cell voltage is reduced. A consequence of the temperature dependency of standard potentials is that the voltage produced by a galvanic cell is also temperature dependent.

Galvanic Corrosion

Galvanic corrosion is a process that degrades metals electrochemically. This corrosion occurs when two dissimilar metals are in contact with each other in the presence of an electrolyte, such as salt water, forming a galvanic cell with H_2 formation on the more noble metal. The resulting electrochemical potential then develops an electric current that electrolytically dissolves the less noble material. A concentration cell can be formed if the same metal is exposed to two different concentrations of electrolyte.

Daniell Cell

Daniell cells.

The Daniell cell is a type of electrochemical cell invented in 1836 by John Frederic Daniell, a British chemist and meteorologist, and consists of a copper pot filled with a copper (II) sulfate solution, in which is immersed an unglazed earthenware container filled with sulfuric acid and a zinc electrode. He was searching for a way to eliminate the hydrogen bubble problem found in the voltaic pile, and his solution was to use a second electrolyte to consume the hydrogen produced by the first. Zinc sulfate may be substituted for the sulfuric acid. The Daniell cell was a great improvement over the existing technology used in the early days of battery development. A later variant of the Daniell cell called the gravity cell or crowfoot cell was invented in the 1860s by a Frenchman named Callaud and became a popular choice for electrical telegraphy.

The Daniell cell is also the historical basis for the contemporary definition of the volt, which is the unit of electromotive force in the International System of Units. The definitions of electrical units that were proposed at the 1881 International Conference of Electricians were designed so that the electromotive force of the Daniell cell would be about 1.0 volts. With contemporary definitions, the standard potential of the Daniell cell at 25 °C is actually 1.10 V.

In the Daniell cell, copper and zinc electrodes are immersed in a solution of copper(II) sulfate and zinc sulfate, respectively. At the anode (negative electrode), zinc is oxidized per the following half reaction:

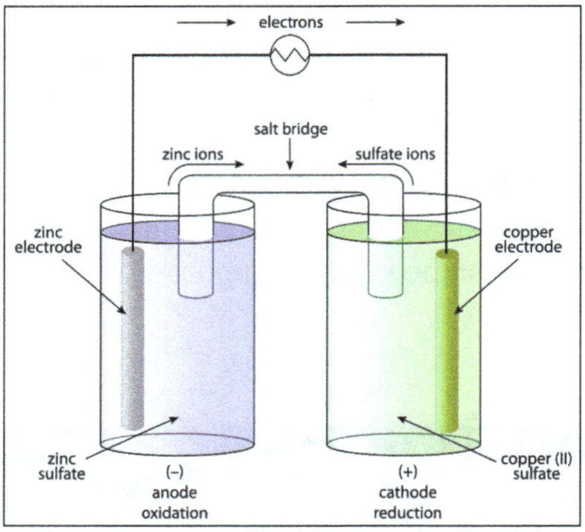

The two half-cell form of the Daniell cell.

$$Zn_{(s)} \rightarrow Zn^{2+}_{(aq)} + 2e^{-} .. \text{(Standard electrode reduction potential} - 0.7618\text{ V)}$$

At the cathode (positive electrode), copper is reduced per the following reaction:

$$Cu^{2+}_{(aq)} + 2e^{-} \rightarrow Cu_{(s)} .. \left(\text{Standard electrode reduction potential} + 0.340\text{ V}\right)$$

Note that positively charged copper ions move towards the positive electrode, driven by a reduction in chemical energy.

The total reaction is:

$$Zn(s) + Cu^{2+}(aq) \rightarrow Zn^{2+}(aq) + Cu(s) .. \left(\text{Open} - \text{circuit voltage} 1.1018\text{ V}\right)$$

These processes result in the accumulation of solid copper at the cathode and the corrosion of the zinc electrode into the solution as zinc cations. The Daniell cell produces approximately 213 kJ per mole (65 g) of zinc. This energy can mostly be attributed to the 207 kJ/mol weaker bonding (smaller magnitude of the cohesive energy) in zinc compared to copper metal, which can be explained in terms of the lack of bonding via partially filled d-orbitals in zinc.

In classroom demonstrations, a form of the Daniell cell known as two half cells is often used due to its simplicity. The two half cells each support one half of the reactions described above. A wire and light bulb may connect the two electrodes. Excess electrons produced by the oxidation of zinc metal are "pushed" out of the anode, which is therefore the negative electrode, travel through the wire and are "pulled" into the copper cathode where they are consumed by the reduction of copper ions. This provides an electrical current that illuminates the bulb.

Since neither half reaction will occur independently of the other, the two half cells must be connected in a way that will allow ions to move freely between them. A porous barrier or ceramic disk may be used to separate the two solutions while allowing the flow of sulfate ions. When the half cells are placed in two entirely different and separate containers, a salt bridge is often used to connect the two cells. The salt bridge typically contains a high concentration of potassium nitrate (a salt that will not interfere chemically with the reaction in either half-cell). In the above wet-cell during discharge, nitrate anions in the salt bridge move into the zinc half-cell in order to balance the increase in Zn^{2+} ions. At the same time, potassium ions from the salt bridge move into the copper half-cell in order to replace the Cu^{2+} ions being precipitated onto the copper electrode.

In the Daniell cell, the porous barrier cannot prevent the flow of copper ions into the zinc half-cell if it is driven by an applied voltage. Hence, recharging (reversing the current flow by an external source of EMF) is impossible because, if the zinc electrode is made to become the cathode, copper ions, rather than zinc ions, will be discharged on account of their lower potential.

Use in Electrometallurgy

Bird's Cell

A variant of the Daniell cell was invented in 1837 by the Guy's hospital physician Golding Bird who used a plaster of Paris barrier to keep the solutions separate. Bird's experiments with this cell were of some importance to the new discipline of electrometallurgy, but Bird himself did not pursue this field; his interest was in electrotherapy. A surprising result from Bird's experiments was the deposition of copper on the porous plaster and in veins running through it without any contact with the metal electrodes. So surprising, in fact, that it was at first disbelieved by electrochemical investigators, including Michael Faraday. Bird himself had to carefully examine his apparatus for inadvertent contact, perhaps through the growth of copper "whiskers", before he was convinced of the result. Deposition of copper, and other metals, had been previously noted, but always previously it had been metal on metal electrode.

Electrotyping

John Dancer, a Liverpool instrument maker, in 1838 was the first to take commercial advantage of the unique features of the Daniell cell for copper plating. In a process now known as electrotyping he found he could make objects to any desired shape by using the porous barrier as a mould. Many others, however, had made the same discovery and in a patent dispute with Thomas Spencer it was pointed out that Bird had priority for the principle. Credit for invention of electrotyping is usually given to the Russian Moritz von Jacobi.

Primary Cell

A variety of standard sizes of primary cells. From left: 4.5V multicell battery, D, C, AA, AAA, AAAA, A23, 9V multicell battery, *(top)* LR44, *(bottom)* CR2032.

A primary cell is a battery (a galvanic cell) that is designed to be used once and discarded, and not recharged with electricity and reused like a secondary cell (rechargeable battery). In general, the electrochemical reaction occurring in the cell is not reversible, rendering the cell unrechargeable. As a primary cell is used, chemical reactions in the battery use up the chemicals that generate the power; when they are gone, the battery stops producing electricity and is useless. In contrast, in a secondary cell, the reaction can be reversed by running a current into the cell with a battery charger to recharge it, regenerating the chemical reactants. Primary cells are made in a range of standard sizes to power small household appliances such as flashlights and portable radios.

Primary batteries make up about 90% of the $50 billion battery market, but secondary batteries have been gaining market share. About 15 billion primary batteries are thrown away worldwide every year, virtually all ending up in landfills. Due to the toxic heavy metals and strong acids they contain, batteries are hazardous waste. Most municipalities classify them as such and require separate disposal. The energy needed to manufacture a battery is about 50 times greater than the energy it contains. Due to their high pollutant content compared to their small energy content, the primary battery is considered a wasteful, environmentally unfriendly technology. Due mainly to increasing sales of wireless devices and cordless tools which cannot be economically powered by primary batteries and come with integral rechargeable batteries, the secondary battery industry has high growth and has slowly been replacing the primary battery in high end products.

Comparison between Primary and Secondary Cells

Secondary cells (rechargeable batteries) are in general more economical to use than primary cells. Their initially higher cost and the purchase cost of a charging system can be spread out over many use cycles (between 100 and 1000 cycles); for example, in hand-held power tools, it would be very costly to replace a high-capacity primary battery pack every few hours of use.

Primary cells are not designed for recharging between manufacturing and use, thus have battery chemistry that has to have a much lower self-discharge rate than older types of secondary cells; but they have lost that advantage with the development of rechargeable secondary cells with very low self discharge rates like low self-discharge NiMH cells that hold enough charge for long enough to be sold as pre-charged.

Common types of secondary cells (namely NiMH and Li-ion) due to their much lower internal resistance do not suffer the large loss of capacity that alkaline, zinc–carbon and zinc chloride ("heavy duty" or "super heavy duty") do with high current draw.

Reserve batteries achieve very long storage time (on the order of 10 years or more) without loss of capacity, by physically separating the components of the battery and only assembling them at the time of use. Such constructions are expensive but are found in applications like munitions, which may be stored for years before use.

Recharging Primary Cells

Attempts to extend the life of primary cells, for example by recharging alkaline batteries, meet with variable results. Third-party devices are manufactured and represented as being able to recharge primary cells. The internal chemical reactions of a primary cell are not easily reversed by externally applied currents, so reactants do not entirely return to their initial state and location. Recharged primary cells will not have the life or performance of secondary cells. Manufacturers of primary cells usually warn against recharging.

Polarization

A major factor reducing the lifetime of primary cells is that they become polarized during use. This means that hydrogen accumulates at the cathode and reduces the effectiveness of the cell. To reduce the effects of polarization in commercial cells and to extend their lives, chemical depolarization is used; that is, an oxidizing agent is added to the cell, to oxidize the hydrogen to water. Manganese dioxide is used in the Leclanché cell and zinc–carbon cell, and nitric acid is used in the Bunsen cell and Grove cell.

Attempts have been made to make simple cells self-depolarizing by roughening the surface of the copper plate to facilitate the detachment of hydrogen bubbles with little success. Electrochemical depolarization exchanges the hydrogen for a metal, such as copper (e.g. Daniell cell), or silver (e.g. silver-oxide cell), so called.

Anode and Cathode

The battery terminal (electrode) that develops a positive voltage polarity (the carbon electrode in a dry cell) is called the cathode and the electrode with a negative polarity (zinc in a dry cell) is called the anode. This is the reverse of the terminology used in an electrolytic cell or thermionic vacuum tube. The reason is that the terms anode and cathode are defined by the direction of electric current, not by their voltage. The anode is the terminal through which conventional current (positive charge) enters the cell from the external circuit, while the cathode is the terminal through which conventional current leaves the cell and flows into the external circuit. Since a battery is a power source which provides the voltage which forces the current through the external circuit, the voltage on the cathode must be higher than the voltage on the anode, creating an electric field directed from cathode to anode, to force the positive charge out of the cathode through the resistance of the external circuit.

Inside the cell the anode is the electrode where chemical oxidation occurs, as it donates electrons which flow out of the anode into the external circuit. The cathode is the electrode where chemical reduction occurs, as it accepts electrons from the circuit through the cathode.

Outside the cell, different terminology is used. As the anode donates positive charge to the electrolyte (thus remaining with an excess of electrons that it will donate to the circuit), it becomes negatively charged and is therefore connected to the terminal marked "–" on the outside of the cell. The cathode, meanwhile, donates negative charge to the electrolyte, so it becomes positively charged (which allows it to accept electrons from the circuit) and is therefore connected to the terminal marked "+" on the outside of the cell.

Secondary Cell

A rechargeable lithium polymer mobile phone battery.

A rechargeable battery, storage battery, secondary cell, or accumulator is a type of electrical battery which can be charged, discharged into a load, and recharged many times, as opposed to a disposable or primary battery, which is supplied fully charged and discarded after use. It is composed of one or more electrochemical cells. The term "accumulator" is used as it accumulates and stores energy through a reversible electrochemical reaction. Rechargeable batteries are produced in many different shapes and sizes, ranging from button cells to megawatt systems connected to stabilize an electrical distribution network. Several different combinations of electrode materials and electrolytes are used, including lead–acid, nickel–cadmium (NiCd), nickel–metal hydride (NiMH), lithium-ion (Li-ion), and lithium-ion polymer (Li-ion polymer).

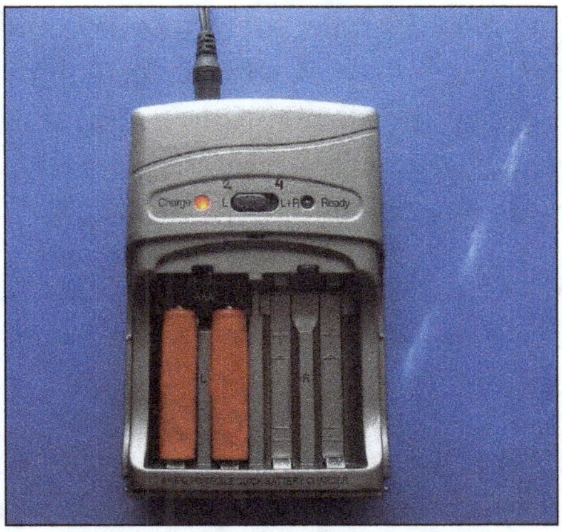

A common consumer battery charger for rechargeable AA and AAA batteries.

Rechargeable batteries typically initially cost more than disposable batteries, but have a much lower total cost of ownership and environmental impact, as they can be recharged inexpensively many times before they need replacing. Some rechargeable battery types are available in the same sizes and voltages as disposable types, and can be used interchangeably with them.

Applications

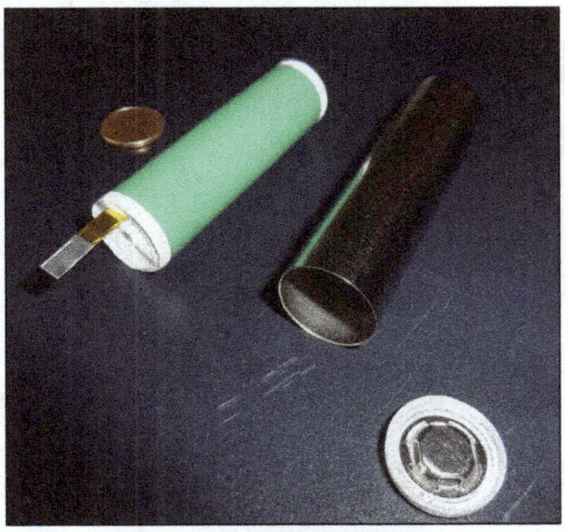

Cylindrical cell (18650) prior to assembly. Several thousand of them (lithium ion) form the Tesla Model S battery.

Lithium ion battery monitoring electronics (over- and discharge protection).

Devices which use rechargeable batteries include automobile starters, portable consumer devices, light vehicles (such as motorized wheelchairs, golf carts, electric bicycles, and electric forklifts), tools, uninterruptible power supplies, and battery storage power stations. Emerging applications in hybrid internal combustion-battery and electric vehicles drive the technology to reduce cost, weight, and size, and increase lifetime.

Older rechargeable batteries self-discharge relatively rapidly, and require charging before first use; some newer low self-discharge NiMH batteries hold their charge for many months, and are typically sold factory-charged to about 70% of their rated capacity.

Battery storage power stations use rechargeable batteries for load-leveling (storing electric energy at times of low demand for use during peak periods) and for renewable energy uses (such as storing power generated from photovoltaic arrays during the day to be used at night). Load-leveling reduces the maximum power which a plant must be able to generate, reducing capital cost and the need for peaking power plants.

According to a report from Research and Markets, the analysts forecast the global rechargeable battery market to grow at a CAGR of 8.32% during the period 2018-2022.

Small rechargeable batteries can power portable electronic devices, power tools, appliances, and so on. Heavy-duty batteries power electric vehicles, ranging from scooters to locomotives and ships. They are used in distributed electricity generation and in stand-alone power systems.

Charging and Discharging

A solar-powered charger for rechargeable AA batteries.

During charging, the positive active material is oxidized, producing electrons, and the negative material is reduced, consuming electrons. These electrons constitute the current flow in the external circuit. The electrolyte may serve as a simple buffer for internal ion flow between the electrodes, as in lithium-ion and nickel-cadmium cells, or it may be an active participant in the electrochemical reaction, as in lead–acid cells.

The energy used to charge rechargeable batteries usually comes from a battery charger using AC mains electricity, although some are equipped to use a vehicle's 12-volt DC power outlet. The voltage of the source must be higher than that of the battery to force current to flow into it, but not too much higher or the battery may be damaged.

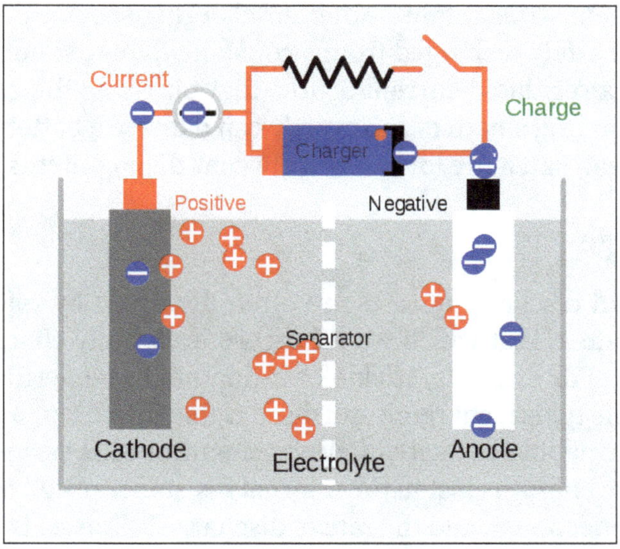

Diagram of the charging of a secondary cell or battery.

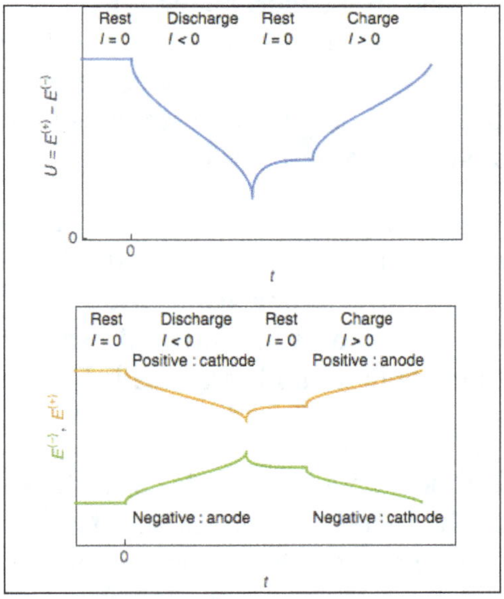

Positive and negative electrode vs. anode and cathode for a secondary battery.

Chargers take from a few minutes to several hours to charge a battery. Slow "dumb" chargers without voltage or temperature-sensing capabilities will charge at a low rate, typically taking 14 hours or more to reach a full charge. Rapid chargers can typically charge cells in two to five hours, depending on the model, with the fastest taking as little as fifteen minutes. Fast chargers must have multiple ways of detecting when a cell reaches full charge (change in terminal voltage, temperature, etc.) to stop charging before harmful overcharging or overheating occurs. The fastest chargers often incorporate cooling fans to keep the cells from overheating. Battery packs intended for rapid charging may include a temperature sensor that the charger uses to protect the pack; the sensor will have one or more additional electrical contacts.

Different battery chemistries require different charging schemes. For example, some battery types can be safely recharged from a constant voltage source. Other types need to be charged with a regulated current source that tapers as the battery reaches fully charged voltage. Charging a battery incorrectly can damage a battery; in extreme cases, batteries can overheat, catch fire, or explosively vent their contents.

Rate of Discharge

Battery charging and discharging rates are often discussed by referencing a "C" rate of current. The C rate is that which would theoretically fully charge or discharge the battery in one hour. For example, trickle charging might be performed at C/20 (or a "20 hour" rate), while typical charging and discharging may occur at C/2 (two hours for full capacity). The available capacity of electrochemical cells varies depending on the discharge rate. Some energy is lost in the internal resistance of cell components (plates, electrolyte, interconnections), and the rate of discharge is limited by the speed at which chemicals in the cell can move about. For lead-acid cells, the relationship between time and discharge rate is described by Peukert's law; a lead-acid cell that can no longer sustain a usable terminal voltage at a high current may still have usable capacity, if discharged at a much lower rate. Data sheets for rechargeable cells often list the discharge capacity on 8-hour or 20-hour or other stated time; cells for uninterruptible power supply systems may be rated at 15 minute discharge.

The terminal voltage of the battery is not constant during charging and discharging. Some types have relatively constant voltage during discharge over much of their capacity. Non-rechargeable alkaline and zinc–carbon cells output 1.5V when new, but this voltage drops with use. Most NiMH AA and AAA cells are rated at 1.2 V, but have a flatter discharge curve than alkalines and can usually be used in equipment designed to use alkaline batteries.

Battery manufacturers' technical notes often refer to voltage per cell (VPC) for the individual cells that make up the battery. For example, to charge a 12 V lead-acid battery (containing 6 cells of 2 V each) at 2.3 VPC requires a voltage of 13.8 V across the battery's terminals.

Damage from Cell Reversal

Subjecting a discharged cell to a current in the direction which tends to discharge it further to the point the positive and negative terminals switch polarity causes a condition called cell reversal. Generally, pushing current through a discharged cell in this way causes undesirable and irreversible chemical reactions to occur, resulting in permanent damage to the cell. Cell reversal can occur under a number of circumstances, the two most common being:

- When a battery or cell is connected to a charging circuit the wrong way around.
- When a battery made of several cells connected in series is deeply discharged.

In the latter case, the problem occurs due to the different cells in a battery having slightly different capacities. When one cell reaches discharge level ahead of the rest, the remaining cells will force the current through the discharged cell.

Many battery-operated devices have a low-voltage cutoff that prevents deep discharges from occurring that might cause cell reversal. A smart battery has voltage monitoring circuitry built inside.

Cell reversal can occur to a weakly charged cell even before it is fully discharged. If the battery drain current is high enough, the cell's internal resistance can create a resistive voltage drop that is greater than the cell's forward emf. This results in the reversal of the cell's polarity while the current is flowing. The higher the required discharge rate of a battery, the better matched the cells should be, both in the type of cell and state of charge, in order to reduce the chances of cell reversal.

In some situations, such as when correcting NiCd batteries that have been previously overcharged, it may be desirable to fully discharge a battery. To avoid damage from the cell reversal effect, it is necessary to access each cell separately: each cell is individually discharged by connecting a load clip across the terminals of each cell, thereby avoiding cell reversal.

Damage during Storage in Fully Discharged State

If a multi-cell battery is fully discharged, it will often be damaged due to the cell reversal effect. It is possible however to fully discharge a battery without causing cell reversal—either by discharging each cell separately, or by allowing each cell's internal leakage to dissipate its charge over time.

Even if a cell is brought to a fully discharged state without reversal, however, damage may occur over time simply due to remaining in the discharged state. An example of this is the sulfation that occurs in lead-acid batteries that are left sitting on a shelf for long periods. For this reason it is often recommended to charge a battery that is intended to remain in storage, and to maintain its charge level by periodically recharging it. Since damage may also occur if the battery is overcharged, the optimal level of charge during storage is typically around 30% to 70%.

Depth of Discharge

Depth of discharge (DOD) is normally stated as a percentage of the nominal ampere-hour capacity; 0% DOD means no discharge. As the usable capacity of a battery system depends on the rate of discharge and the allowable voltage at the end of discharge, the depth of discharge must be qualified to show the way it is to be measured. Due to variations during manufacture and aging, the DOD for complete discharge can change over time or number of charge cycles. Generally a rechargeable battery system will tolerate more charge/discharge cycles if the DOD is lower on each cycle.

Lifespan and Cycle Stability

If batteries are used repeatedly even without mistreatment, they lose capacity as the number of charge cycles increases, until they are eventually considered to have reached the end of their useful life. Different battery systems have differing mechanisms for wearing out. For example, in lead-acid batteries, not all the active material is restored to the plates on each charge/discharge cycle; eventually enough material is lost that the battery capacity is reduced. In lithium-ion types, especially on deep discharge, some reactive lithium metal can be formed on charging, which is no longer available to participate in the next discharge cycle. Sealed batteries may lose moisture from their liquid electrolyte, especially if overcharged or operated at high temperature. This reduces the cycling life.

Recharging Time

BYD e6 taxi. Recharging in 15 Minutes to 80 percent.

Recharging time is an important parameter to the user of a product powered by rechargeable batteries. Even if the charging power supply provides enough power to operate the device as well as recharge the battery, the device is attached to an external power supply during the charging time. For electric vehicles used industrially, charging during off-shifts may be acceptable. For highway electric vehicles, rapid charging is necessary for charging in a reasonable time.

A rechargeable battery cannot be recharged at an arbitrarily high rate. The internal resistance of the battery will produce heat, and excessive temperature rise will damage or destroy a battery. For some types, the maximum charging rate will be limited by the speed at which active material can diffuse through a liquid electrolyte. High charging rates may produce excess gas in a battery, or may result in damaging side reactions that permanently lower the battery capacity. Very roughly, and with many exceptions and details, restoring a battery's full capacity in one hour or less is considered fast charging. A battery charger system will include more complex control-circuit- and charging strategies for fast charging, than for a charger designed for slower recharging.

Active Components

The active components in a secondary cell are the chemicals that make up the positive and negative active materials, and the electrolyte. The positive and negative are made up of different materials, with the positive exhibiting a reduction potential and the negative having an oxidation potential. The sum of these potentials is the standard cell potential or voltage.

In primary cells the positive and negative electrodes are known as the cathode and anode, respectively. Although this convention is sometimes carried through to rechargeable systems—especially with lithium-ion cells, because of their origins in primary lithium cells—this practice can lead to confusion. In rechargeable cells the positive electrode is the cathode on discharge and the anode on charge, and vice versa for the negative electrode.

Types

The lead–acid battery, invented in 1859 by French physicist Gaston Planté, is the oldest type of rechargeable battery. Despite having a very low energy-to-weight ratio and a low energy-to-volume ratio, its ability to supply high surge currents means that the cells have a relatively large power-to-weight ratio. These features, along with the low cost, makes it attractive for use in motor vehicles to provide the high current required by automobile starter motors.

The nickel–cadmium battery (NiCd) was invented by Waldemar Jungner of Sweden in 1899. It uses nickel oxide hydroxide and metallic cadmium as electrodes. Cadmium is a toxic element, and was banned for most uses by the European Union in 2004. Nickel–cadmium batteries have been almost completely superseded by nickel–metal hydride (NiMH) batteries.

The nickel–metal hydride battery (NiMH) became available in 1989. These are now a common consumer and industrial type. The battery has a hydrogen-absorbing alloy for the negative electrode instead of cadmium.

The lithium-ion battery was introduced in the market in 1991, is the choice in most consumer electronics, having the best energy density and a very slow loss of charge when not in use. It does have drawbacks too, particularly the risk of unexpected ignition from the heat generated by the battery. Such incidents are rare and according to experts, they can be minimized "via appropriate design, installation, procedures and layers of safeguards" so the risk is acceptable.

Lithium-ion polymer batteries (LiPo) are light in weight, offer slightly higher energy density than Li-ion at slightly higher cost, and can be made in any shape. They are available but have not displaced Li-ion in the market. A primary use is for LiPo batteries is in powering remote-controlled cars, boats and airplanes. LiPo packs are readily available on the consumer market, in various configurations, up to 44.4v, for powering certain R/C vehicles and helicopters or drones. Some test reports warn of the risk of fire when the batteries are not used in accordance with the instructions. Independent reviews of the technology discuss the risk of fire and explosion from Lithium-ion batteries under certain conditions because they use liquid electrolytes.

Other Experimental Types

Type	Voltage[a]	Energy density[b]			Power[c]	E/$[e]	Self-disch.[f]	Charge Efficiency	Cycles[g]	Life[h]
	(V)	(MJ/kg)	(Wh/kg)	(Wh/L)	(W/kg)	(Wh/$)	(%/month)	(%)	(#)	(years)
Lithium sulfur	2.0	0.94-1.44	400	350					~1400	
Sodium-ion	3.6			30		3.3			5000+	Testing
Thin film lithium	?		300	959	6000	?[p]			40000	
Zinc-bromide	1.8	0.27-0.31	75-85							
Zinc-cerium	2.5									Under testing
Vanadium redox	1.15-1.55	0.09-0.13	25-35				20%		20,000	25 years
Sodium-sulfur		0.54	150					89–92%	2500–4500	
Molten salt	2.58	0.25-1.04	70-290	160	150-220	4.54			3000+	<=20
Silver-zinc	1.86	0.47	130	240						
Quantum Battery (oxide semiconductor)	1.5-3			500	8000(W/L)				100,000	

In the table:

- [a] Nominal cell voltage in V.
- [b] Energy density = energy/weight or energy/size, given in three different units.
- [c] Specific power = power/weight in W/kg.
- [e] Energy/consumer price in W·h/US$ (approximately).
- [f] Self-discharge rate in %/month.
- [g] Cycle durability in number of cycles.
- [h] Time durability in years.
- [i] VRLA or recombinant includes gel batteries and absorbed glass mats.
- [p] Pilot production.

The lithium–sulfur battery was developed by Sion Power in 1994. The company claims superior energy density to other lithium technologies.

The thin film battery (TFB) is a refinement of lithium ion technology by Excellatron. The developers claim a large increase in recharge cycles to around 40,000 and higher charge and discharge rates, at least 5 C charge rate. Sustained 60 C discharge and 1000C peak discharge rate and a significant increase in specific energy, and energy density.

UltraBattery, a hybrid lead-acid battery and ultracapacitor invented by Australia's national science organisation CSIRO, exhibits tens of thousands of partial state of charge cycles and has outperformed traditional lead-acid, lithium and NiMH-based cells when compared in testing in this mode against variability management power profiles. UltraBattery has kW and MW-scale installations in place in Australia, Japan and the U.S.A. It has also been subjected to extensive testing in hybrid electric vehicles and has been shown to last more than 100,000 vehicle miles in on-road commercial testing in a courier vehicle. The technology is claimed to have a lifetime of 7 to 10 times that of conventional lead-acid batteries in high rate partial state-of-charge use, with safety and environmental benefits claimed over competitors like lithium-ion. Its manufacturer suggests an almost 100% recycling rate is already in place for the product.

The potassium-ion battery delivers around a million cycles, due to the extraordinary electrochemical stability of potassium insertion/extraction materials such as Prussian blue.

The sodium-ion battery is meant for stationary storage and competes with lead–acid batteries. It aims at a low total cost of ownership per kWh of storage. This is achieved by a long and stable lifetime. The effective number of cycles is above 5000 and the battery is not damaged by deep discharge. The energy density is rather low, somewhat lower than lead–acid.

Alternatives

A rechargeable battery is only one of several types of rechargeable energy storage systems. Several alternatives to rechargeable batteries exist or are under development. For uses such as portable radios, rechargeable batteries may be replaced by clockwork mechanisms which are wound up by hand, driving dynamos, although this system may be used to charge a battery rather than to operate the radio directly. Flashlights may be driven by a dynamo directly. For transportation, uninterruptible power supply systems and laboratories, flywheel energy storage systems store energy in a spinning rotor for conversion to electric power when needed; such systems may be used to provide large pulses of power that would otherwise be objectionable on a common electrical grid.

Ultracapacitors—capacitors of extremely high value— are also used; an electric screwdriver which charges in 90 seconds and will drive about half as many screws as a device using a rechargeable battery was introduced in 2007, and similar flashlights have been produced. In keeping with the concept of ultracapacitors, betavoltaic batteries may be utilized as a method of providing a trickle-charge to a secondary battery, greatly extending the life and energy capacity of the battery system being employed; this type of arrangement is often referred to as a "hybrid betavoltaic power source" by those in the industry.

Ultracapacitors are being developed for transportation, using a large capacitor to store energy instead of the rechargeable battery banks used in hybrid vehicles. One drawback of capacitors compared to batteries is that the terminal voltage drops rapidly; a capacitor that has 25% of its initial energy left in it will have one-half of its initial voltage. By contrast, battery systems tend to have a terminal voltage that does not decline rapidly until nearly exhausted. The undesirable characteristic complicates the design of power electronics for use with ultracapacitors. However, there are potential benefits in cycle efficiency, lifetime, and weight compared with rechargeable systems. China started using ultracapacitors on two commercial bus routes in 2006; one of them is route 11 in Shanghai.

Flow batteries, used for specialized applications, are recharged by replacing the electrolyte liquid. A flow battery can be considered to be a type of rechargeable fuel cell.

Electrolysis

In chemistry and manufacturing, electrolysis is a technique that uses a direct electric current (DC) to drive an otherwise non-spontaneous chemical reaction. Electrolysis is commercially important as a stage in the separation of elements from naturally occurring sources such as ores using an electrolytic cell. The voltage that is needed for electrolysis to occur is called the decomposition potential.

Electrolysis is the passing of a direct electric current through an ionic substance that is either molten or dissolved in a suitable solvent, producing chemical reactions at the electrodes and decomposition of the materials.

The main components required to achieve electrolysis are:

- An electrolyte: a substance, frequently an ion-conducting polymer that contains free ions, which carry electric current in the electrolyte. If the ions are not mobile, as in most solid salts, then electrolysis cannot occur.
- A direct current (DC) electrical supply: provides the energy necessary to create or discharge the ions in the electrolyte. Electric current is carried by electrons in the external circuit.

Electrodes of metal, graphite and semiconductor material are widely used. Choice of suitable electrode depends on chemical reactivity between the electrode and electrolyte and manufacturing cost.

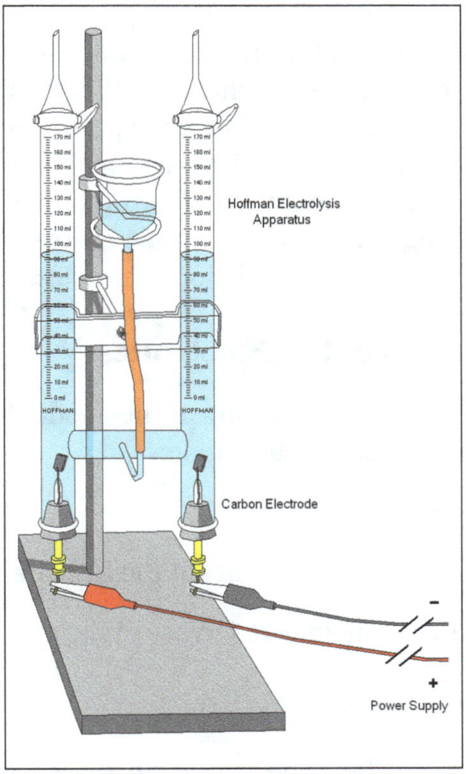

Illustration of an electrolysis apparatus used in a school laboratory.

Process of Electrolysis

The key process of electrolysis is the interchange of atoms and ions by the removal or addition of electrons from the external circuit. The desired products of electrolysis are often in a different physical state from the electrolyte and can be removed by some physical processes. For example, in the electrolysis of brine to produce hydrogen and chlorine, the products are gaseous. These gaseous products bubble from the electrolyte and are collected.

$$2\,NaCl + 2\,H_2O \rightarrow 2\,NaOH + H_2 + Cl$$

A liquid containing electrolyte is produced by:

- Solvation or reaction of an ionic compound with a solvent (such as water) to produce mobile ions.
- An ionic compound is melted by heating.

An electrical potential is applied across a pair of electrodes immersed in the electrolyte.

Each electrode attracts ions that are of the opposite charge. Positively charged ions (cations) move towards the electron-providing (negative) cathode. Negatively charged ions (anions) move towards the electron-extracting (positive) anode.

In this process electrons are either absorbed or released. Neutral atoms gain or lose electrons and become charged ions that then pass into the electrolyte. The formation of uncharged atoms from ions is called discharging. When an ion gains or loses enough electrons to become uncharged (neutral) atoms, the newly formed atoms separate from the electrolyte. Positive metal ions like Cu^{2+}deposit onto the cathode in a layer. The terms for this are electroplating, electrowinning, and electrorefining. When an ion gains or loses electrons without becoming neutral, its electronic charge is altered in the process. In chemistry, the loss of electrons is called oxidation, while electron gain is called reduction.

Oxidation and Reduction at the Electrodes

Oxidation of ions or neutral molecules occurs at the anode. For example, it is possible to oxidize ferrous ions to ferric ions at the anode:

$$Fe^{2+}(aq) \rightarrow Fe^{3+}(aq) + e^{-}$$

Reduction of ions or neutral molecules occurs at the cathode.

It is possible to reduce ferricyanide ions to ferrocyanide ions at the cathode:

$$Fe(CN)_{6}^{3-} + e^{-} \rightarrow Fe(CN)_{6}^{4-}$$

Neutral molecules can also react at either of the electrodes. For example: p-Benzoquinone can be reduced to hydroquinone at the cathode:

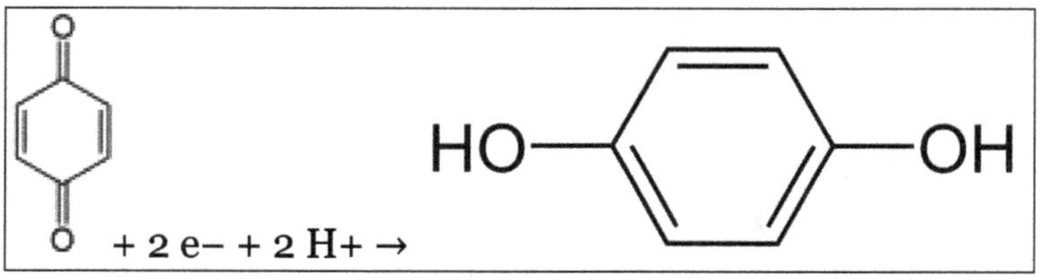

In the last example, H^{+} ions (hydrogen ions) also take part in the reaction and are provided by the acid in the solution, or by the solvent itself (water, methanol, etc.). Electrolysis reactions involving H^{+} ions are fairly common in acidic solutions. In aqueous alkaline solutions, reactions involving OH^{-} (hydroxide ions) are common.

Sometimes the solvents themselves (usually water) are oxidized or reduced at the electrodes. It is even possible to have electrolysis involving gases. Such as when using a Gas diffusion electrode.

Energy Changes during Electrolysis

The amount of electrical energy that must be added equals the change in Gibbs free energy of the reaction plus the losses in the system. The losses can (in theory) be arbitrarily close to zero, so the maximum thermodynamic efficiency equals the enthalpy change divided by the free energy change of the reaction. In most cases, the electric input is larger than the enthalpy change of the reaction, so some energy is released in the form of heat. In some cases, for instance, in the electrolysis of steam into hydrogen and oxygen at high temperature, the opposite is true and heat energy is absorbed. This heat is absorbed from the surroundings, and the heating value of the produced hydrogen is higher than the electric input.

Variations

Pulsating current results in products different from DC. For example, pulsing increases the ratio of ozone to oxygen produced at the anode in the electrolysis of an aqueous acidic solution such as dilute sulphuric acid. Electrolysis of ethanol with pulsed current evolves an aldehyde instead of primarily an acid.

Related Techniques

The following techniques are related to electrolysis:

- Electrochemical cells, including the hydrogen fuel cell, use differences in Standard electrode potential to generate an electrical potential that provides useful power. Though related via the interaction of ions and electrolysis and the operation of electrochemical cells are quite distinct. However, a chemical cell should not be seen as performing electrolysis in reverse.

Faraday's Laws of Electrolysis

First Law of Electrolysis

In 1832, Michael Faraday reported that the quantity of elements separated by passing an electric current through a molten or dissolved salt is proportional to the quantity of electric charge passed through the circuit.

This became the basis of the first law of electrolysis. The mass of the substance (m) deposited or liberated at an electrode is directly proportional to the quantity of electricity or charge (Q) passed. In this equation k is equal to the electromechanical constant.

$$m = k{\cdot}q$$

or

$$m = e \cdot Q$$

where; e is known as the electrochemical equivalent of the metal deposited or of the gas liberated at the electrode.

Second Law of Electrolysis

Faraday discovered that when the same amount of current is passed through different electrolytes/elements connected in series, the mass of substance liberated/deposited at the electrodes is directly proportional to their equivalent weight.

Industrial Uses

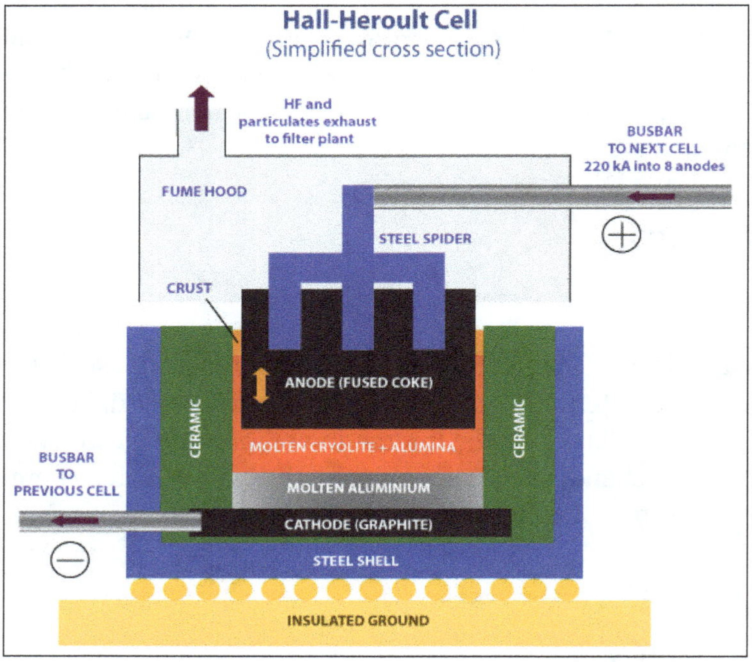

Hall-Heroult process for producing aluminium.

- Electrometallurgy is the process of reduction of metals from metallic compounds to obtain the pure form of metal using electrolysis. Aluminium, lithium, sodium, potassium, magnesium, calcium, and in some cases copper, are produced in this way.
- Production of chlorine and sodium hydroxide.
- Production of sodium chlorate and potassium chlorate.
- Production of perfluorinated organic compounds such as trifluoroacetic acid by the process of electrofluorination.
- Production of electrolytic copper as a cathode, from refined copper of lower purity as an anode.

Electrolysis has many other uses:

- Production of oxygen for spacecraft and nuclear submarines.
- Production of hydrogen for fuel, using a cheap source of electrical energy.

Electrolysis is also used in the cleaning and preservation of old artifacts. Because the process separates the non-metallic particles from the metallic ones, it is very useful for cleaning a wide variety of metallic objects, from old coins to even larger objects including rusted cast iron cylinder blocks and heads when rebuilding automobile engines. Rust removal from small iron or steel objects by electrolysis can be done in a home workshop using simple materials such as a plastic bucket, tap water, lengths of rebar, washing soda, baling wire, and a battery charger.

Manufacturing Processes

In manufacturing, electrolysis can be used for:

- Electroplating, where a thin film of metal is deposited over a substrate material. Electroplating is used in many industries for either functional or decorative purposes, as in-vehicle bodies and nickel coins.
- Electrochemical machining (ECM), where an electrolytic cathode is used as a shaped tool for removing material by anodic oxidation from a workpiece. ECM is often used as a technique for deburring or for etching metal surfaces like tools or knives with a permanent mark or logo.

Competing Half-reactions in Solution Electrolysis

Using a cell containing inert platinum electrodes, electrolysis of aqueous solutions of some salts leads to the reduction of the cations (e.g., metal deposition with, e.g., zinc salts) and oxidation of the anions (e.g. evolution of bromine with bromides). However, with salts of some metals (e.g. sodium) hydrogen is evolved at the cathode, and for salts containing some anions (e.g. sulfate SO_4^{2-}) oxygen is evolved at the anode. In both cases, this is due to water being reduced to form hydrogen or oxidized to form oxygen. In principle, the voltage required to electrolyze a salt solution can be derived from the standard electrode potential for the reactions at the anode and cathode. The standard electrode potential is directly related to the Gibbs free energy, ΔG, for the reactions at each electrode and refers to an electrode with no current flowing. An extract from the table of standard electrode potentials is shown below.

Half-reaction	E° (V)
$Na^{+} + e^{-} \rightleftharpoons Na(s)$	−2.71
$Zn^{2+} + 2e^{-} \rightleftharpoons Zn(s)$	−0.7618
$2H^{+} + 2e^{-} \rightleftharpoons H_2(g)$	≡ 0
$Br_2(aq) + 2e^{-} \rightleftharpoons 2Br^{-}$	+1.0873

$O_2(g) + 4H^+ + 4e^- \rightleftharpoons 2H_2O$	+1.23
$Cl_2(g) + 2e^- \rightleftharpoons 2Cl^-$	+1.36
$S_2O_8^{2-} + 2e^- \rightleftharpoons 2SO_4^{2-}$	+2.07

In terms of electrolysis, this table should be interpreted as follows:

- Oxidized species (often a cation) with a more negative cell potential are more difficult to reduce than oxidized species with a more positive cell potential. For example, it is more difficult to reduce a sodium ion to a sodium metal than it is to reduce a zinc ion to a zinc metal.
- Reduced species (often an anion) with a more positive cell potential are more difficult to oxidize than reduced species with a more negative cell potential. For example, it is more difficult to oxidize sulfate anions than it is to oxidize bromide anions.

Using the Nernst equation the electrode potential can be calculated for a specific concentration of ions, temperature and the number of electrons involved. For pure water (pH 7):

- The electrode potential for the reduction producing hydrogen is −0.41 V.
- The electrode potential for the oxidation producing oxygen is +0.82 V.

Comparable figures calculated in a similar way, for 1M zinc bromide, $ZnBr_2$, are −0.76 V for the reduction to Zn metal and +1.10 V for the oxidation producing bromine. The conclusion from these figures is that hydrogen should be produced at the cathode and oxygen at the anode from the electrolysis of water—which is at variance with the experimental observation that zinc metal is deposited and bromine is produced. The explanation is that these calculated potentials only indicate the thermodynamically preferred reaction. In practice, many other factors have to be taken into account such as the kinetics of some of the reaction steps involved. These factors together mean that a higher potential is required for the reduction and oxidation of water than predicted, and these are termed overpotentials. Experimentally it is known that overpotentials depend on the design of the cell and the nature of the electrodes.

For the electrolysis of a neutral (pH 7) sodium chloride solution, the reduction of sodium ion is thermodynamically very difficult and water is reduced evolving hydrogen leaving hydroxide ions in solution. At the anode the oxidation of chlorine is observed rather than the oxidation of water since the overpotential for the oxidation of chloride to chlorine is lower than the overpotential for the oxidation of water to oxygen. The hydroxide ions and dissolved chlorine gas react further to form hypochlorous acid. The aqueous solutions resulting from this process is called electrolyzed water and is used as a disinfectant and cleaning agent.

Applications of Electrochemistry

Electrochemistry has a number of different uses, particularly in industry. The principles of cells are used to make electrical batteries. In science and technology, a battery is a device that stores chemical energy and makes it available in an electrical form. Batteries are made of electrochemical devices such as one or more galvanic cells or fuel cells. Batteries have many uses including in:

- Torches.
- Electrical appliances such as cellphones (long-life alkaline batteries).
- Digital cameras (lithium batteries).
- Hearing aids (silver-oxide batteries).
- Digital watches (mercury/silver-oxide batteries).
- Military applications (thermal batteries).

We are going to look at a few examples of the uses of electrochemistry in industry.

Electroplating (ESCRQ)

The electrolytic cell can be used for electroplating.

Electroplating occurs when an electrically conductive object is coated with a layer of metal using electrical current. Sometimes, electroplating is used to give a metal particular properties or for aesthetic reasons:

- Corrosion protection.
- Abrasion and wear resistance.
- The production of jewellery.

a) An electroplated piece of aluminium artwork and
b) a wax stool electroplated in copper.

Electro-refining (also sometimes called electrowinning) is electroplating on a large scale. Copper plays a major role in the electrical industry as it is very conductive and is used in electric cables. One of the problems though is that copper must be pure if it is to be an effective current carrier. One of the methods used to purify copper, is electrowinning (copper ore is processed into impure *blister copper*, which is then deposited as pure copper through electroplating). The copper electrowinning process is as follows:

- A bar of impure copper containing other metallic impurities acts as the anode.
- The cathode is made up of pure copper with few impurities.
- The electrolyte is a solution of aqueous $CuSO_4$ and H_2SO_4.

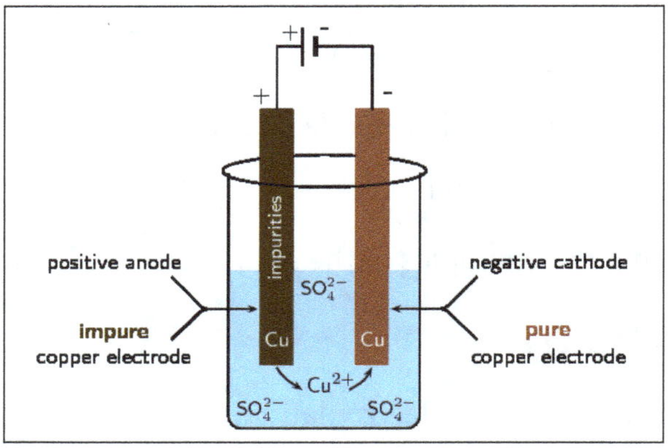

A simplified diagram to illustrate what happens during the electrowinning of copper.

- When current passes through the cell, electrolysis takes place.

  The impure copper anode oxidises to form Cu2+ ions in solution. The anode decreases in mass. $Cu(s) \rightarrow Cu^{2+}(aq) + 2e^{-}$.

  At the cathode reduction of positive copper ions takes place to produce pure copper metal. The cathode increases in mass. $Cu^{2+}(aq) + 2e^{-} \rightarrow Cu(s)$ (> 99% purity).

- The other metal impurities do not dissolve Au(s), Ag(s)) and form a solid sludge at the bottom of the tank or remain in solution (Zn(aq), Fe(aq) and Pb(aq)) in the electrolyte.

The Chloralkali Industry (ESCRR)

The chlorine-alkali (chloralkali) industry is an important part of the chemical industry, which produces chlorine and sodium hydroxide through the electrolysis of the raw material brine. Brine is a saturated solution of sodium chloride (NaCl) that is obtained from natural salt deposits.

The products of the chloralkali industry have a number of important uses.

Chlorine is used:

- To purify water.
- As a disinfectant.
- In the production of:
    - Hypochlorous acid (used to kill bacteria in drinking water).
    - Paper, food.
    - Antiseptics, insecticides, medicines, textiles, laboratory chemicals.
    - Paints, petroleum products, solvents, plastics (such as polyvinyl chloride).

Sodium hydroxide (also known as 'caustic soda') is used to:

- Make soap and other cleaning agents.
- Purify bauxite (the ore of aluminium).
- Make paper.
- Make rayon (artificial silk).

One of the problems of producing chlorine and sodium hydroxide is that when they are produced together the chlorine combines with the sodium hydroxide to form chlorate (ClO^-) and chloride (Cl^-) ions. This leads to the production of sodium chlorate, NaClO, a component of household bleach.

To overcome this problem the chlorine and sodium hydroxide must be separated from each other so that they don't react. There are three industrial processes that have been designed to overcome this problem. All three methods involve electrolytic cells.

- The anode is a carbon electrode suspended from the top of a chamber.
- The cathode is liquid mercury that flows along the floor of this chamber.
- The electrolyte is brine (NaCl solution) that is passed through the chamber.
- When an electric current is applied to the circuit, chloride ions in the electrolyte are oxidised at the anode to form chlorine gas.

$$2Cl^-(aq) \rightarrow Cl_2(g) + 2e^-$$

- At the same time sodium ions are reduced at the anode to solid sodium. The solid sodium dissolves in the mercury making a sodium/mercury amalgam.

$$Na^+(aq) + Hg(l) + e^- \rightarrow Na(Hg)$$

- The amalgam is poured into a separate vessel, where it decomposes into sodium and mercury.
- The sodium reacts with water in the vessel and produces sodium hydroxide and hydrogen gas, while the mercury returns to the electrolytic cell to be used again.

$$2\,Na(Hg) + 2\,H_2O(l) \rightarrow 2\,NaOH(aq) + H_2(g) + Hg(l)$$

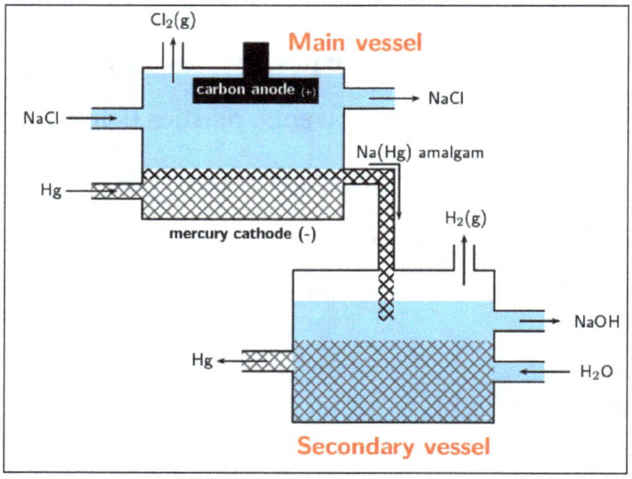

The mercury cell.

- This method only produces a fraction of the chlorine and sodium hydroxide that is used by industry as it has certain disadvantages:
    - Mercury is expensive and toxic.
    - Some mercury always escapes with the brine that has been used.
    - Mercury reacts with the brine to form mercury(II) chloride.
    - The mercury cell requires a lot of electricity.
    - Although the chlorine gas produced is very pure, mercury has to be removed from the sodium hydroxide and hydrogen gas mixture.

In the past the effluent was released into lakes and rivers, causing mercury to accumulate in fish and other animals feeding on the fish. Today, the brine is treated before it is discharged so that the environmental impact is lower.

In the diaphragm-cell:

- A porous diaphragm divides the electrolytic cell into an anode compartment and a cathode compartment.
- Brine is introduced into the anode compartment and flows through the diaphragm into the cathode compartment.

- An electric current is passed through the brine causing the salt's chlorine ions and sodium ions to move to the electrodes.
- Chlorine gas is produced at the anode.

  $$2\,Cl^{-}(aq) + 2e^{-} \rightarrow Cl_2(g)$$

- At the cathode, sodium ions react with water forming caustic soda (NaOH) and hydrogen gas.

  $$2\,Na^{+}(aq) + 2\,H_2O(l) + e^{-} \rightarrow 2\,NaOH(aq) + H_2(g)$$

- Some NaCl salt remains in the solution with the caustic soda and can be removed at a later stage.

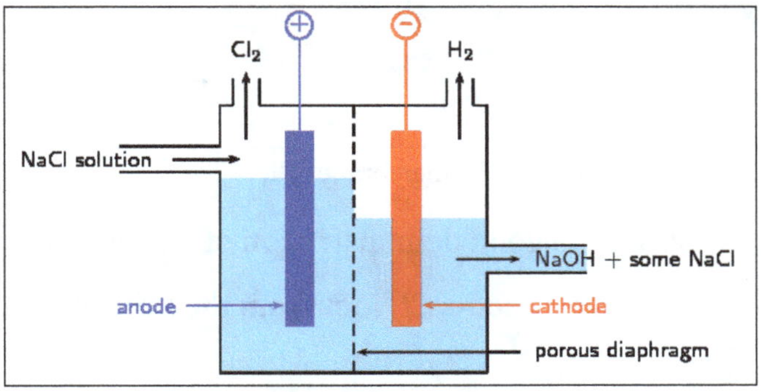

The diaphragm cell.

The following animation gives a good demonstration of how a diaphragm cell works.

The advantages of the diaphragm cell are:

- Uses less energy than the mercury cell.
- Does not contain toxic mercury.

It also has disadvantages however:

- The sodium hydroxide is much less concentrated and not as pure.
- The chlorine gas often contains oxygen gas as well.
- The process is less cost-effective as the sodium hydroxide solution needs to be concentrated and purified before it can be used.

The Membrane Cell

The membrane cell is very similar to the diaphragm cell, and the same reactions occur. The main differences are:

- The two electrodes are separated by an ion-selective membrane, rather than by a diaphragm.

- The membrane structure allows cations to pass through it between compartments of the cell but does not allow anions to pass through (this has nothing to do with the size of the pores, but rather with the charge on the ions).
- Brine is pumped into the anode compartment, and only the positively charged sodium ions pass into the cathode compartment, which contains pure water.

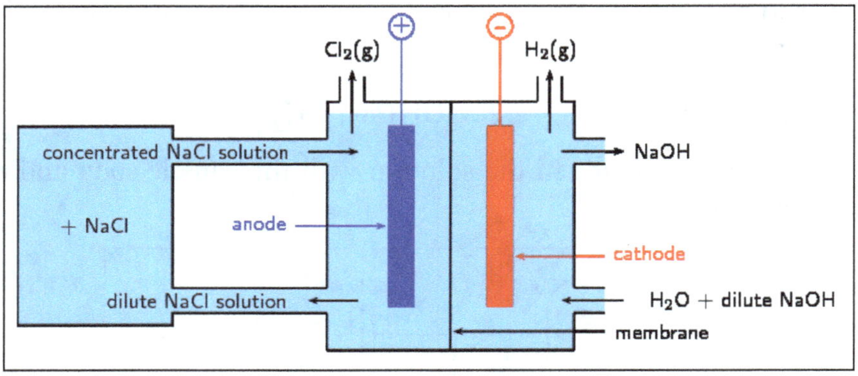

The membrane cell.

The following animation gives a good demonstration of how a membrane cell works.

- At the positively charged anode, Cl^- ions from the brine are oxidised to Cl_2 gas.

$$2Cl^-(aq) \rightarrow Cl_2(g) + 2e^-$$

- At the negatively charged cathode, hydrogen ions in the water are reduced to hydrogen gas.

$$2H_2O(l) + 2e^- \rightarrow H_2(g) + 2OH^-$$

- The Na^+ ions flow through the membrane to the cathode compartment and react with the remaining hydroxide (OH^-) ions from the water to form sodium hydroxide (NaOH).

$$Na+(aq) + OH-(aq) \rightarrow NaOH(aq)$$

- The chloride ions cannot pass through the membrane, so the chlorine does not come into contact with the sodium hydroxide in the cathode compartment. The sodium hydroxide is removed from the cell. The overall equation is as follows:

$$2NaCl(aq) + 2H_2O(l) \rightarrow Cl_2(g) + H_2(g) + 2NaOH(g)$$

The advantages of using this method are:

- The sodium hydroxide that is produced is very pure because it is kept separate from the sodium chloride solution.

- The sodium hydroxide has a relatively high concentration.
- This process uses the least electricity of all three cells.
- The cell is cheaper to operate than the other two cells.
- The cell does not contain toxic mercury or asbestos.

References

- Basics-of-Electrochemistry, Electrochemistry, Supplemental-Modules-(Analytical-Chemistry), Analytical-Chemistry: chem.libretexts.org, Retrieved 17 May, 2020
- Schmidt-Rohr, K. (2018). "How Batteries Store and Release Energy: Explaining Basic Electrochemistry" "J. Chem. Educ." 95: 1801-1810. Https://doi.org/10.1021/acs.jchemed.8b00479
- Electrochemical-reaction, science: britannica.com, Retrieved 19 April, 2020
- David Linden, Thomas B. Reddy (ed). Handbook Of Batteries 3rd Edition. Mcgraw-Hill, New York, 2002 ISBN 0-07-135978-8
- IUPAC, Compendium of Chemical Terminology, 2nd ed. (the "Gold Book") (1997). Online corrected version: (2006–) "Electrochemical potential". Doi:10.1351/goldbook.E01945
- Hill, Marquita K. (2004). Understanding Environmental Pollution: A Primer. Cambridge University Press. P. 274. ISBN 0521527260

Chemical Thermodynamics 4

- Thermodynamic System
- Spontaneous Process
- Nonspontaneous Process
- The Laws of Thermodynamics
- Hess's Law
- Internal Energy
- Enthalpy
- Entropy
- Gibbs Free Energy

The study of the interrelation of heat and work with chemical reactions or with physical changes of state within the laws of thermodynamics is known as chemical thermodynamics. Internal energy, enthalpy, entropy, Gibbs free energy, etc. are some of the concepts that fall under its domain. This chapter closely examines these key concepts of chemical thermodynamics to provide an extensive understanding of the subject.

Chemical thermodynamics is the study of how heat and work relate to each other both in changes of state and in chemical reactions. It involves a series of rules and laws that explain how heat and work, well, work, and explains which processes can happen spontaneously and which need some help.

There are several basic principles of chemical thermodynamics to consider: systems, the laws of thermodynamics, and enthalpy. Chemical thermodynamics is also concerned with four particular quantities: Internal energy, enthalpy, entropy, and the Gibbs free energy.

Chemical thermodynamics allows us to predict both the direction and extent of spontaneous chemical and physical change under particular conditions. There are 4 important thermodynamic functions:

- Internal energy (U): The sum of all the energies of all the individual particles in a sample of matter. $\Delta U = U_{nal} - U_{initial} = q + w$
- Enthalpy (H): Function related to the heat absorbed or evolved by a chemical system during a chemical reaction or physical change under conditions of constant pressure.
- Entropy (S): A measure of the number of ways energy is distributed throughout a chemical system.
- Gibbs energy (G): G = H – TS, where T is the thermodynamic temperature, measured in kelvin.

Thermodynamic Concepts

Heat and Temperature

- Heat (q): a transfer of energy across a boundary due to a temperature difference.
- Heat lows spontaneously from a hot object to a cold object until both are the same.
- temperature (i.e.2 bodies are at thermal equilibrium).
- Exothermic process: q<0 (heat lows out of the system, energy released).
- Endothermic process: q>0 (heats lows into the system, energy is absorbed).
- Thermal equilibrium: q=0 (no net heat low).
- Temperature is a measurement of internal energy of a substance.

System, Surroundings and Universe

- System: that part of the universe under study (usually 1 or more chemical species) and separated from the surroundings by a real or imaginary boundary. A system may be open, closed or isolated.
- Open systems: can gain or lose mass and energy across boundaries (e.g. human body).
- Closed systems: can absorb or release energy but not mass across the boundary (e.g. light bulb- light and heat can transfer).
- Isolated systems: cant exchange ma!er or energy with their surroundings. (e.g. stoppered vacuum flask).

- Processes that occur in an isolated system are adiabatic (not passable).

- Universe: system + surroundings.

Units

- SI units for energy, work (motion against an opposing force) and heat is joules (J).
    - $1\,J = 1 kgm^2s^{-2}$
    - Since joules is not much energy we use kJ.
    - Calories are also used (cal): 1 cal=4.184J.
- $\Delta G = \Delta H - T\Delta S$ (convert all to joules or kilojoules).
- For all variables it's always final- initial e.g. $\Delta T = T_{final} - T_{initial}$ and $\Delta G = G_{products} - G_{reactants}$
- For chemical reactions, we denote this using the symbols $\Delta_r U$, $\Delta_r H$, $\Delta_r S$ and $\Delta_r G$.

The structure of chemical thermodynamics is based on the first two laws of thermodynamics. Starting from the first and second laws of thermodynamics, four equations called the "fundamental equations of Gibbs" can be derived. From these four, a multitude of equations, relating the thermodynamic properties of the thermodynamic system can be derived using relatively simple mathematics. This outlines the mathematical framework of chemical thermodynamics.

The primary objective of chemical thermodynamics is the establishment of a criterion for determination of the feasibility or spontaneity of a given transformation. In this manner, chemical thermodynamics is typically used to predict the energy exchanges that occur in the following processes:

- Chemical reactions.
- Phase changes.
- The formation of solutions.

The following state functions are of primary concern in chemical thermodynamics:

- Internal energy (U).
- Enthalpy (H).
- Entropy (S).
- Gibbs free energy (G).

Most identities in chemical thermodynamics arise from application of the first and second laws of thermodynamics, particularly the law of conservation of energy, to these state functions.

The 3 Laws of Thermodynamics

- The energy of the universe is constant.
- In any spontaneous process, there is always an increase in entropy of the universe.
- The entropy of a perfect crystal (well ordered) at 0 Kelvin is zero.

Chemical Energy

Chemical energy is the potential of a chemical substance to undergo a transformation through a chemical reaction or to transform other chemical substances. Breaking or making of chemical bonds involves energy or heat, which may be either absorbed or evolved from a chemical system.

Energy that can be released (or absorbed) because of a reaction between a set of chemical substances is equal to the difference between the energy content of the products and the reactants. This change in energy is called the change in internal energy of a chemical reaction. Where $\Delta_f U^{\circ}_{reactants}$ is the internal energy of formation of the reactant molecules that can be calculated from the bond energies of the various chemical bonds of the molecules under consideration and $\Delta_f U^{\circ}_{reactants}$ is the internal energy of formation of the product molecules. The change in internal energy is a process which is equal to the heat change if it is measured under conditions of constant volume (at STP condition), as in a closed rigid container such as a bomb calorimeter. However, under conditions of constant pressure, as in reactions in vessels open to the atmosphere, the measured heat change is not always equal to the internal energy change, because pressure-volume work also releases or absorbs energy. (The heat change at constant pressure is called the enthalpy change; in this case the enthalpy of formation).

Another useful term is the heat of combustion, which is the energy released due to a combustion reaction and often applied in the study of fuels. Food is similar to hydrocarbon fuel and carbohydrate fuels, and when it is oxidized, its caloric content is similar (though not assessed in the same way as a hydrocarbon fuel).

In chemical thermodynamics the term used for the chemical potential energy is chemical potential, and for chemical transformation an equation most often used is the Gibbs-Duhem equation.

Chemical Reactions

In most cases of interest in chemical thermodynamics there are internal degrees of freedom and processes, such as chemical reactions and phase transitions, which always create entropy unless they are at equilibrium, or are maintained at a "running equilibrium" through "quasi-static" changes by being coupled to constraining devices, such as pistons or electrodes, to deliver and receive external work. Even for homogeneous "bulk" materials, the free energy functions depend on the composition, as do all the extensive thermodynamic potentials, including the internal energy. If the quantities $\{N_i\}$, the number of chemical species, are omitted from the formulae, it is impossible to describe compositional changes.

Gibbs Function or Gibbs Energy

For a "bulk" (unstructured) system they are the last remaining extensive variables. For an unstructured, homogeneous "bulk" system, there are still various extensive compositional variables $\{N_i\}$ that G depends on, which specify the composition (the amounts of each chemical substance, expressed as the numbers of molecules present or the numbers of moles). Explicitly,

$$G = G(T, P, \{N_i\}).$$

For the case where only PV work is possible,

$$dG = -SdT + VdP + \sum_i \mu_i dN_i$$

in which μ_i is the chemical potential for the i-th component in the system,

$$\mu_i = \left(\frac{\partial G}{\partial N_i}\right)_{T,P,N_{j\neq i},\text{etc.}}.$$

The expression for dG is especially useful at constant T and P, conditions which are easy to achieve experimentally and which approximates the condition in living creatures,

$$(dG)_{T,P} = \sum_i \mu_i dN_i.$$

Chemical Affinity

While this formulation is mathematically defensible, it is not particularly transparent since one does not simply add or remove molecules from a system. There is always a process involved in changing the composition; e.g., a chemical reaction (or many), or movement of molecules from one phase (liquid) to another (gas or solid). We should find a notation which does not seem to imply that the amounts of the components (N_i) can be changed independently. All real processes obey conservation of mass, and in addition, conservation of the numbers of atoms of each kind. Whatever molecules are transferred to or from should be considered part of the "system".

Consequently, we introduce an explicit variable to represent the degree of advancement of a process, a progress variable ξ for the extent of reaction and to the use of the partial derivative $\partial G/\partial \xi$ (in place of the widely used "ΔG", since the quantity at issue is not a finite change). The result is an understandable expression for the dependence of dG on chemical reactions (or other processes). If there is just one reaction,

$$(dG)_{T,P} = \left(\frac{\partial G}{\partial \xi}\right)_{T,P} d\xi.$$

If we introduce the stoichiometric coefficient for the i-th component in the reaction,

$$\nu_i = \partial N_i / \partial \xi$$

which tells how many molecules of i are produced or consumed, we obtain an algebraic expression for the partial derivative,

$$\left(\frac{\partial G}{\partial \xi}\right)_{T,P} = \sum_i \mu_i \nu_i = -\mathbb{A}$$

where, we introduce a concise and historical name for this quantity, the "affinity", symbolized by A, as introduced by Théophile de Donder in 1923. The minus sign comes from the fact the affinity was defined to represent the rule that spontaneous changes will ensue only when the change in the Gibbs free energy of the process is negative, meaning that the chemical species have a positive affinity for each other. The differential for G takes on a simple form which displays its dependence on compositional change,

$$(dG)_{T,P} = -\mathbb{A}\, d\xi.$$

If there are a number of chemical reactions going on simultaneously, as is usually the case,

$$(dG)_{T,P} = -\sum_k \mathbb{A}_k\, d\xi_k.$$

a set of reaction coordinates $\{\xi_j\}$, avoiding the notion that the amounts of the components (N_i) can be changed independently. The expressions above are equal to zero at thermodynamic equilibrium, while in the general case for real systems, they are negative because all chemical reactions proceeding at a finite rate produce entropy. This can be made even more explicit by introducing the reaction rates $d\xi_j/dt$. For each and every physically independent process.

$$\mathbb{A}\,\dot{\xi} \leq 0.$$

This is a remarkable result since the chemical potentials are intensive system variables, depending only on the local molecular milieu. They cannot "know" whether the temperature and pressure (or any other system variables) are going to be held constant over time. It is a purely local criterion and must hold regardless of any such constraints. Of course, it could have been obtained by taking partial derivatives of any of the other fundamental state functions, but nonetheless is a general criterion for (−T times) the entropy production from that spontaneous process; or at least any part of it that is not captured as external work.

We now relax the requirement of a homogeneous "bulk" system by letting the chemical potentials and the affinity apply to any locality in which a chemical reaction (or any other process) is occurring. By accounting for the entropy production due to irreversible processes, the inequality for dG is now replaced by an equality,

$$dG = -SdT + VdP - \sum_k \mathbb{A}_k \, d\xi_k + W'$$

or

$$dG_{T,P} = -\sum_k \mathbb{A}_k \, d\xi_k + W'.$$

Any decrease in the Gibbs function of a system is the upper limit for any isothermal, isobaric work that can be captured in the surroundings, or it may simply be dissipated, appearing as T times a corresponding increase in the entropy of the system and/or its surrounding. Or it may go partly toward doing external work and partly toward creating entropy. The important point is that the extent of reaction for a chemical reaction may be coupled to the displacement of some external mechanical or electrical quantity in such a way that one can advance only if the other one also does. The coupling may occasionally be rigid, but it is often flexible and variable.

In solution chemistry and biochemistry, the Gibbs free energy decrease ($\partial G/\partial \xi$, in molar units, denoted cryptically by ΔG) is commonly used as a surrogate for (−T times) the entropy produced by spontaneous chemical reactions in situations where there is no work being done; or at least no "useful" work; i.e., other than perhaps some $\pm$ P dV. The assertion that all spontaneous reactions have a negative ΔG is merely a restatement of the fundamental thermodynamic relation, giving it the physical dimensions of energy and somewhat obscuring its significance in terms of entropy. When there is no useful work being done, it would be less misleading to use the Legendre transforms of the entropy appropriate for constant T, or for constant T and P, the Massieu functions $-F/T$ and $-G/T$ respectively.

Non Equilibrium

Generally the systems treated with the conventional chemical thermodynamics are either at equilibrium or near equilibrium. Ilya Prigogine developed the thermodynamic treatment of open systems that are far from equilibrium. In doing so he has discovered phenomena and structures of completely new and completely unexpected types. His generalized, nonlinear and irreversible thermodynamics has found surprising applications in a wide variety of fields.

The non equilibrium thermodynamics has been applied for explaining how ordered structures e.g. the biological systems, can develop from disorder. Even if Onsager's relations are utilized, the classical principles of equilibrium in thermodynamics still show that linear systems close to equilibrium always develop into states of disorder which are stable to perturbations and cannot explain the occurrence of ordered structures.

Prigogine called these systems dissipative systems, because they are formed and maintained by the dissipative processes which take place because of the exchange of energy between the system and its environment and because they disappear if that exchange ceases. They may be said to live in symbiosis with their environment.

The method which Prigogine used to study the stability of the dissipative structures to perturbations is of very great general interest. It makes it possible to study the most varied problems, such as city traffic problems, the stability of insect communities, the development of ordered biological structures and the growth of cancer cells to mention but a few examples.

System Constraints

In this regard, it is crucial to understand the role of walls and other constraints, and the distinction between independent processes and coupling. Contrary to the clear implications of many reference sources, the previous analysis is not restricted to homogeneous, isotropic bulk systems which can deliver only PdV work to the outside world, but applies even to the most structured systems. There are complex systems with many chemical "reactions" going on at the same time, some of which are really only parts of the same, overall process. An independent process is one that could proceed even if all others were unaccountably stopped in their tracks. Understanding this is perhaps a "thought experiment" in chemical kinetics, but actual examples exist.

A gas reaction which results in an increase in the number of molecules will lead to an increase in volume at constant external pressure. If it occurs inside a cylinder closed with a piston, the equilibrated reaction can proceed only by doing work against an external force on the piston. The extent variable for the reaction can increase only if the piston moves, and conversely, if the piston is pushed inward, the reaction is driven backwards.

Similarly, a redox reaction might occur in an electrochemical cell with the passage of current in wires connecting the electrodes. The half-cell reactions at the electrodes are constrained if no current is allowed to flow. The current might be dissipated as joule heating, or it might in turn run an electrical device like a motor doing mechanical work. An automobile lead-acid battery can be recharged, driving the chemical reaction backwards. In this case as well, the reaction is not an independent process. Some, perhaps most, of the Gibbs free energy of reaction may be delivered as external work.

The hydrolysis of ATP to ADP and phosphate can drive the force times distance work delivered by living muscles, and synthesis of ATP is in turn driven by a redox chain in mitochondria and chloroplasts, which involves the transport of ions across the membranes of these cellular organelles. The coupling of processes here, is often not complete. Gas can leak slowly past a piston, just as it can slowly leak out of a rubber balloon. Some reaction may occur in a battery even if no external current is flowing. There is usually a coupling coefficient, which may depend on relative rates, which determines what percentage of the driving free energy is turned into external work, or captured as "chemical work"; a misnomer for the free energy of another chemical process.

Thermodynamic System

A thermodynamic system includes anything whose thermodynamic properties are of interest. It is embedded in its surroundings or environment; it can exchange heat with, and do work on, its environment through a boundary, which is the imagined wall that separates the system and the environment. In reality, the immediate surroundings of the system are interacting with it directly and therefore have a much stronger influence on its behavior and properties. For example, if we are studying a car engine, the burning gasoline inside the cylinder of the engine is the thermodynamic system; the piston, exhaust system, radiator, and air outside form the surroundings of the system. The boundary then consists of the inner surfaces of the cylinder and piston.

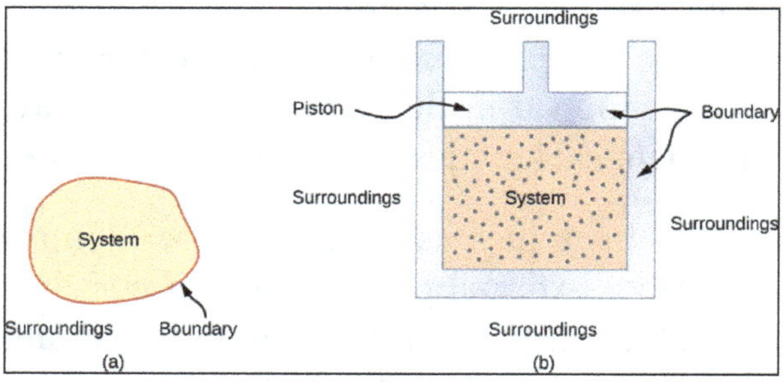

A system, which can include any relevant process or value, is self-contained in an area. The surroundings may also have relevant information; however, the surroundings are important to study only if the situation is an open system. (b) The burning gasoline in the cylinder of a car engine is an example of a thermodynamic system.

Normally, a system must have some interactions with its surroundings. A system is called an isolated or closed system if it is completely separated from its environment—for example, a gas that is surrounded by immovable and thermally insulating walls. In reality, a closed system does not exist unless the entire universe is treated as the system, or it is used as a model for an actual system that has minimal interactions with its environment. Most systems are known as an open system, which can exchange energy and/or matter with its surroundings.

(a) (b)

Figure (a) This boiling tea kettle is an open thermodynamic system. It transfers heat and matter (steam) to its surroundings. (b) A pressure cooker is a good approximation to a closed system. A little steam escapes through the top valve to prevent explosion.

When we examine a thermodynamic system, we ignore the difference in behavior from place to place inside the system for a given moment. In other words, we concentrate on the macroscopic properties of the system, which are the averages of the microscopic properties of all the molecules or entities in the system. Any thermodynamic system is therefore treated as a continuum that has the same behavior everywhere inside. We assume the system is in equilibrium. You could have, for example, a temperature gradient across the system.

Before we can carry out any study on a thermodynamic system, we need a fundamental characterization of the system. When we studied a mechanical system, we focused on the forces and torques on the system, and their balances dictated the mechanical equilibrium of the system. In a similar way, we should examine the heat transfer between a thermodynamic system and its environment or between the different parts of the system, and its balance should dictate the thermal equilibrium of the system. Intuitively, such a balance is reached if the temperature becomes the same for different objects or parts of the system in thermal contact, and the net heat transfer over time becomes zero.

Thus, when we say two objects (a thermodynamic system and its environment, for example) are in thermal equilibrium, we mean that they are at the same temperature. Let us consider three objects at temperatures T_1,T_2, and T_3 respectively. How do we know whether they are in thermal equilibrium? The governing principle here is the zeroth law of thermodynamics:

Mathematically, we can simply write the zeroth law of thermodynamics as

$$If\ T_1 = T_2\ and T_1 = T_3, then\ T_2 = T_3.$$

This is the most fundamental way of defining temperature: Two objects must be at the same temperature thermodynamically if the net heat transfer between them is zero when they are put in thermal contact and have reached a thermal equilibrium.

The zeroth law of thermodynamics is equally applicable to the different parts of a closed system and requires that the temperature everywhere inside the system be the same if the system has reached a thermal equilibrium. To simplify our discussion, we assume the system is uniform with only one type of material—for example, water in a tank. The measurable properties of the system at least include its volume, pressure, and temperature. The range of specific relevant variables depends upon the system. For example, for a stretched rubber band, the relevant variables would be length, tension, and temperature. The relationship between these three basic properties of the system is called the equation of state of the system and is written symbolically for a closed system as

$$f(p,V,T)=0,$$

where V, p, and T are the volume, pressure, and temperature of the system at a given condition.

In principle, this equation of state exists for any thermodynamic system but is not always readily available. The forms of $f(p,V,T)=0$ for many materials have been determined either experimentally or theoretically.

We have so far introduced several physical properties that are relevant to the thermodynamics of a thermodynamic system, such as its volume, pressure, and temperature. We can separate these quantities into two generic categories. The quantity associated with an amount of matter is an extensive variable, such as the volume and the number of moles. The other properties of a system are intensive variables, such as the pressure and temperature. An extensive variable doubles its value if the amount of matter in the system doubles, provided all the intensive variables remain the same. For example, the volume or total energy of the system doubles if we double the amount of matter in the system while holding the temperature and pressure of the system unchanged.

Spontaneous Process

A spontaneous process is the time-evolution of a system in which it releases free energy and it moves to a lower, more thermodynamically stable energy state. The sign convention for free energy follows the general convention for thermodynamic measurements, in which a release of free energy from the system corresponds to a negative change in the free energy of the system and a positive change in the free energy of the surroundings.

Depending on the nature of the process, the free energy is determined differently. For example, the Gibbs free energy is used when considering processes that occur under constant pressure and temperature conditions whereas the Helmholtz free energy is used when considering processes that occur under constant volume and temperature conditions.

Because spontaneous processes are characterized by a decrease in the system's free energy, they do not need to be driven by an outside source of energy.

For cases involving an isolated system where no energy is exchanged with the surroundings, spontaneous processes are characterized by an increase in entropy.

In general, the spontaneity of a process only determines whether or not a process *can* occur and makes no indication as to whether or not the process *will* occur. In other words, spontaneity is a necessary, but not sufficient, condition for a process to actually occur. Furthermore, spontaneity makes no implication as to the speed at which as spontaneous may occur.

As an example, the conversion of a diamond into graphite is a spontaneous process at room temperature and pressure. Despite being spontaneous, this process does not occur since the energy to break the strong carbon-carbon bonds is larger than the release in free energy.

Using Free Energy to Determine Spontaneity

For a process that occurs at constant temperature and pressure, spontaneity can be determined using the change in Gibbs free energy, which is given by:

$$\Delta G = \Delta H - T\Delta S,$$

where the sign of ΔG depends on the signs of the changes in enthalpy (ΔH) and entropy (ΔS). The sign of ΔG will change from positive to negative (or vice versa) where $T = \Delta H/\Delta S$.

In cases where ΔG is:

- Negative, the process is spontaneous and may proceed in the forward direction as written.
- Positive, the process is non-spontaneous as written, but it may proceed spontaneously in the *reverse direction.*
- Zero, the process is at equilibrium, with no net change taking place over time.

This set of rules can be used to determine four distinct cases by examining the signs of the ΔS and ΔH.

- When $\Delta S > 0$ and $\Delta H < 0$, the process is always spontaneous as written.
- When $\Delta S < 0$ and $\Delta H > 0$, the process is never spontaneous, but the reverse process is always spontaneous.
- When $\Delta S > 0$ and $\Delta H > 0$, the process will be spontaneous at high temperatures and non-spontaneous at low temperatures.
- When $\Delta S < 0$ and $\Delta H < 0$, the process will be spontaneous at low temperatures and non-spontaneous at high temperatures.

For the latter two cases, the temperature at which the spontaneity changes will be determined by the relative magnitudes of ΔS and ΔH.

Using Entropy to determine Spontaneity

When using the entropy change of a process to assess spontaneity, it is important to carefully consider the definition of the system and surroundings. The second law of thermodynamics states that a process involving an isolated system will be spontaneous if the entropy of the system increases over time. For open or closed systems, however, the statement must be modified to say that the total entropy of the *combined* system and surroundings must increase, or,

$$\Delta S_{\text{total}} = \Delta S_{\text{system}} + \Delta S_{\text{surroundings}} \geq 0.$$

This criterion can then be used to explain how it is possible for the entropy of an open or closed system to decrease during a spontaneous process. A decrease in system entropy can only occur spontaneously if the entropy change of the surroundings is both positive in sign and has a larger magnitude than the entropy change of the system:

$$\Delta S_{\text{surroundings}} > 0$$

and

$$\left|\Delta S_{\text{surroundings}}\right| > \left|\Delta S_{\text{system}}\right|$$

In many processes, the increase in entropy of the surroundings is accomplished via heat transfer from the system to the surroundings (i.e. an exothermic process).

Nonspontaneous Process

An endergonic reaction (also called a nonspontaneous reaction or an unfavorable reaction) is a chemical reaction in which the standard change in free energy is positive, and energy is absorbed. The total amount of energy is a loss (it takes more energy to start the reaction than what is gotten out of it) so the total energy is a negative net result. Endergonic reactions can also be pushed by coupling them to another reaction, which is strongly exergonic, through a shared intermediate.

The Laws of Thermodynamics

The three laws of thermodynamics define physical quantities (temperature, energy, and entropy) that characterize thermodynamic systems at thermal equilibrium. The laws describe how these quantities behave under various circumstances, and preclude the possibility of certain phenomena (such as perpetual motion).

The three laws of thermodynamics are:

- First law of thermodynamics: When energy passes, as work, as heat, or with matter, into or out of a system, the system's internal energy changes in accord with the law of conservation of energy. Equivalently, perpetual motion machines of the first kind (machines that produce work with no energy input) are impossible.
- Second law of thermodynamics: In a natural thermodynamic process, the sum of the entropies of the interacting thermodynamic systems increases. Equivalently, perpetual motion machines of the second kind (machines that spontaneously convert thermal energy into mechanical work) are impossible.
- Third law of thermodynamics: The entropy of a system approaches a constant value as the temperature approaches absolute zero. With the exception of non-crystalline solids (glasses) the entropy of a system at absolute zero is typically close to zero.

In addition, there is conventionally added a "zeroth law", which defines thermal equilibrium:

- Zeroth law of thermodynamics: If two systems are each in thermal equilibrium with a third system, they are in thermal equilibrium with each other. This law helps define the concept of temperature.

The laws of thermodynamics are important fundamental laws in physics and they are applicable in other natural sciences.

Zeroth Law

The zeroth law of thermodynamics may be stated in the following form:

If two systems are both in thermal equilibrium with a third system then they are in thermal equilibrium with each other.

The law is intended to allow the existence of an empirical parameter, the temperature, as a property of a system such that systems in thermal equilibrium with each other have the same temperature. The law as stated here is compatible with the use of a particular physical body, for example a mass of gas, to match temperatures of other bodies, but does not justify regarding temperature as a quantity that can be measured on a scale of real numbers.

Though this version of the law is one of the most commonly stated versions, it is only one of a diversity of statements that are labeled as "the zeroth law" by competent writers. Some statements go further so as to supply the important physical fact that temperature is one-dimensional and that one can conceptually arrange bodies in real number sequence from colder to hotter. Perhaps there exists no unique "best possible statement" of the "zeroth law", because there is in the literature a range of formulations of the principles of thermodynamics, each of which call for their respectively appropriate versions of the law.

Although these concepts of temperature and of thermal equilibrium are fundamental to thermodynamics and were clearly stated in the nineteenth century, the desire to explicitly number the above law was not widely felt until Fowler and Guggenheim did so in the 1930s, long after the first, second, and third law were already widely understood and recognized. Hence it was numbered the zeroth law. The importance of the law as a foundation to the earlier laws is that it allows the definition of temperature in a non-circular way without reference to entropy, its conjugate variable. Such a temperature definition is said to be 'empirical'.

First Law

The first law of thermodynamics may be stated in several ways :

The increase in internal energy of a closed system is equal to the total of the energy added to the system. In particular, if the energy entering the system is supplied as heat and if energy leaves the system as work, the heat is accounted for as positive and the work as negative.

$$\Delta U_{\text{system}} = Q - W$$

In the case of a thermodynamic cycle of a closed system, which returns to its original state, the heat Q_{in} supplied to the system in one stage of the cycle, minus the heat Q_{out} removed from it in another stage of the cycle, plus the work added to the system W_{in} equals the work that leaves the system W_{out}.

$$\Delta U_{\text{system(full cycle)}} = 0$$

hence, for a full cycle,

$$Q = Q_{in} - Q_{out} + W_{in} - W_{out} = W_{net}$$

For the particular case of a thermally isolated system (adiabatically isolated), the change of the internal energy of an adiabatically isolated system can only be the result of the work added to the system, because the adiabatic assumption is: $Q = 0$.

$$\Delta U_{\text{system}} = U_{\text{final}} - U_{\text{initial}} = W_{in} - W_{out}$$

For processes that include transfer of matter, a further statement is needed: 'With due account of the respective fiducial reference states of the systems, when two systems, which may be of different chemical compositions, initially separated only by an impermeable wall, and otherwise isolated, are combined into a new system by the thermodynamic operation of removal of the wall, then,

$$U_{\text{system}} = U_1 + U_2,$$

where U_{system} denotes the internal energy of the combined system, and U_1 and U_2 denote the internal energies of the respective separated systems.'

More specifically, the First Law encompasses several principles:

- The law of conservation of energy.

  This states that energy can be neither created nor destroyed. However, energy can change forms, and energy can flow from one place to another. A particular consequence of the law of conservation of energy is that the total energy of an isolated system does not change.

- The concept of internal energy and its relationship to temperature.

  If a system has a definite temperature, then its total energy has three distinguishable components. If the system is in motion as a whole, it has kinetic energy. If the system as a whole is in an externally imposed force field (e.g. gravity), it has potential energy relative to some reference point in space. Finally, it has internal energy, which is a fundamental quantity of thermodynamics. The establishment of the concept of internal energy distinguishes the first law of thermodynamics from the more general law of conservation of energy.

  $$E_{\text{total}} = \text{KE}_{\text{system}} + \text{PE}_{\text{system}} + U_{\text{system}}$$

  The internal energy of a substance can be explained as the sum of the diverse kinetic energies of the erratic microscopic motions of its constituent atoms, and of the potential energy of interactions between them. Those microscopic energy terms are collectively called the substance's internal energy (U), and are accounted for by macroscopic thermodynamic property. The total of the kinetic energies of microscopic motions of the constituent atoms increases as the system's temperature increases; this assumes no other interactions at the microscopic level of the system such as chemical reactions, potential energy of constituent atoms with respect to each other.

- Work is a process of transferring energy to or from a system in ways that can be described by macroscopic mechanical forces exerted by factors in the surroundings, outside the system. Examples are an externally driven shaft agitating a stirrer within the system, or an externally imposed electric field that polarizes the material of the system, or a piston that compresses the system. Unless otherwise stated, it is customary to treat work as occurring without its dissipation to the surroundings. Practically speaking, in all natural process, some of the work is dissipated by internal friction or viscosity. The work done by the system can come from its overall kinetic energy, from its overall potential energy, or from its internal energy.

For example, when a machine (not a part of the system) lifts a system upwards, some energy is transferred from the machine to the system. The system's energy increases as work is done on the system and in this particular case, the energy increase of the system is manifested as an increase in the system's gravitational potential energy. Work added to the system increases the Potential Energy of the system:

$$W = \Delta \mathrm{PE}_{\text{system}}$$

Or in general, the energy added to the system in the form of work can be partitioned to kinetic, potential or internal energy forms:

$$W = \Delta \mathrm{KE}_{\text{system}} + \Delta \mathrm{PE}_{\text{system}} + \Delta U_{\text{system}}$$

- When matter is transferred into a system, that masses' associated internal energy and potential energy are transferred with it.

  $$(u \Delta M)_{\text{in}} = \Delta U_{\text{system}}$$

  where u denotes the internal energy per unit mass of the transferred matter, as measured while in the surroundings; and ΔM denotes the amount of transferred mass.

- The flow of heat is a form of energy transfer.

  Heating is a natural process of moving energy to or from a system other than by work or the transfer of matter. Direct passage of heat is only from a hotter to a colder system.

  If the system has rigid walls that are impermeable to matter, and consequently energy cannot be transferred as work into or out from the system, and no external long-range force field affects it that could change its internal energy, then the internal energy can only be changed by the transfer of energy as heat:

  $$\Delta U_{\text{system}} = Q$$

where Q denotes the amount of energy transferred into the system as heat.

Combining these principles leads to one traditional statement of the first law of thermodynamics: it is not possible to construct a machine which will perpetually output work without an equal amount of energy input to that machine. Or more briefly, a perpetual motion machine of the first kind is impossible.

Second Law

The second law of thermodynamics indicates the irreversibility of natural processes, and, in many cases, the tendency of natural processes to lead towards spatial homogeneity of matter and energy, and especially of temperature. It can be formulated in a variety of interesting and important ways.

It implies the existence of a quantity called the entropy of a thermodynamic system. In terms of this quantity it implies that, when two initially isolated systems in separate but nearby regions of space, each in thermodynamic equilibrium with itself but not necessarily with each other, are then allowed to interact, they will eventually reach a mutual thermodynamic equilibrium. The sum of the entropies of the initially isolated systems is less than or equal to the total entropy of the final combination. Equality occurs just when the two original systems have all their respective intensive variables (temperature, pressure) equal; then the final system also has the same values.

The second law is applicable to a wide variety of processes, reversible and irreversible. All natural processes are irreversible. Reversible processes are a useful and convenient theoretical fiction, but do not occur in nature.

A prime example of irreversibility is in the transfer of heat by conduction or radiation. It was known long before the discovery of the notion of entropy that when two bodies initially of different temperatures come into thermal connection, then heat always flows from the hotter body to the colder one.

The second law tells also about kinds of irreversibility other than heat transfer, for example those of friction and viscosity, and those of chemical reactions. The notion of entropy is needed to provide that wider scope of the law.

According to the second law of thermodynamics, in a theoretical and fictive reversible heat transfer, an element of heat transferred, δQ, is the product of the temperature (T), both of the system and of the sources or destination of the heat, with the increment (dS) of the system's conjugate variable, its entropy (S),

$$Q \quad \delta Q = T\,dS$$

Entropy may also be viewed as a physical measure of the lack of physical information about the microscopic details of the motion and configuration of a system, when only the macroscopic states are known. This lack of information is often described as *disorder* on a microscopic or molecular scale. The law asserts that for two given macroscopically specified states of a system, there is a quantity called the difference of information entropy between them. This information entropy difference defines how much additional microscopic physical information is needed to specify one of the macroscopically specified states, given the macroscopic specification of the other – often a conveniently chosen reference state which may be presupposed to exist rather than explicitly stated. A final condition of a natural process always contains microscopically specifiable effects which are not fully and exactly predictable from the macroscopic specification of the initial condition of the process. This is why entropy increases in natural processes – the increase tells how much extra microscopic information is needed to distinguish the final macroscopically specified state from the initial macroscopically specified state.

Third Law

The third law of thermodynamics is sometimes stated as follows:

> The entropy of a perfect crystal of any pure substance approaches zero as the temperature approaches absolute zero.

At zero temperature the system must be in a state with the minimum thermal energy. This statement holds true if the perfect crystal has only one state with minimum energy. Entropy is related to the number of possible microstates according to:

$$S = k_B \ln \Omega$$

Where S is the entropy of the system, k_B Boltzmann's constant, and Ω the number of microstates (e.g. possible configurations of atoms). At absolute zero there is only 1 microstate possible (Ω=1 as all the atoms are identical for a pure substance and as a result all orders are identical as there is only one combination) and ln(1) = 0.

A more general form of the third law that applies to a system such as a glass that may have more than one minimum microscopically distinct energy state, or may have a microscopically distinct state that is "frozen in" though not a strictly minimum energy state and not strictly speaking a state of thermodynamic equilibrium, at absolute zero temperature:

> The entropy of a system approaches a constant value as the temperature approaches zero.

The constant value (not necessarily zero) is called the residual entropy of the system.

Hess's Law

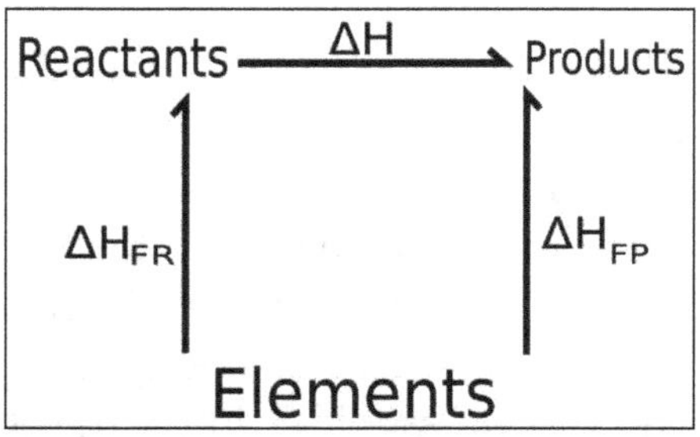

A representation of Hess' law (where H represents enthalpy).

Hess' law of constant heat summation, also known as Hess' law (or Hess's law), is a relationship in physical chemistry named after Germain Hess, a Switzerland-born Russian chemist and physician who published it in 1840. The law states that the total enthalpy change during the complete course of a chemical reaction is the same whether the reaction is made in one step or in several steps.

Hess' law is now understood as an expression of the principle of conservation of energy, also expressed in the first law of thermodynamics, and the fact that the enthalpy of a chemical process is independent of the path taken from the initial to the final state (i.e. enthalpy is a state function). Reaction enthalpy changes can be determined by calorimetry for many reactions. The values are usually stated for processes with the same initial and final temperatures and pressures, although the conditions can vary during the reaction. Hess' law can be used to determine the overall energy required for a chemical reaction, when it can be divided into synthetic steps that are individually easier to characterize. This affords the compilation of standard enthalpies of formation, that may be used as a basis to design complex.

Hess' law states that the change of enthalpy in a chemical reaction (i.e. the heat of reaction at constant pressure) is independent of the pathway between the initial and final states.

In other words, if a chemical change takes place by several different routes, the overall enthalpy change is the same, regardless of the route by which the chemical change occurs (provided the initial and final condition are the same).

Hess' law allows the enthalpy change (ΔH) for a reaction to be calculated even when it cannot be measured directly. This is accomplished by performing basic algebraic operations based on the chemical equations of reactions using previously determined values for the enthalpies of formation.

Addition of chemical equations leads to a net or overall equation. If enthalpy change is known for each equation, the result will be the enthalpy change for the net equation. If the net enthalpy change is negative ($\Delta H_{net} < 0$), the reaction is exothermic and is more likely to be spontaneous; positive ΔH values correspond to endothermic reactions. Entropy also plays an important role in determining spontaneity, as some reactions with a positive enthalpy change are nevertheless spontaneous.

Hess' law states that enthalpy changes are additive. Thus the ΔH for a single reaction,

$$\Delta H^{\ominus}_{reaction} = \sum \Delta H^{\ominus}_{\mathrm{f}(products)} - \sum \Delta H^{\ominus}_{\mathrm{f}(reactants)}$$

where ΔH_f is an enthalpy of formation, and the $\ominus$ superscript indicates standard state values. This may be considered as the sum of two (real or fictitious) reactions:

Reactants → Elements

$$\Delta H^{\ominus} = -\sum \Delta H^{\ominus}_{\mathrm{f}(reactants)}$$

and Elements → Products

$$\Delta H^{\ominus} = \sum \Delta H^{\ominus}_{\mathrm{f}(products)}$$

Examples

- Given:
  - $B_2O_3(s)+3H_2O(g)\rightarrow 3O_2(g)+B_2H_6(g)(\Delta H=2035\ kJ/mol)$
  - $H_2O(l)\rightarrow H_2O(g)\ (\Delta H = 44\ kJ/mol)$
  - $H_2(g)+(1/2)O_2(g)\rightarrow H_2O(l)(\Delta H = -286\ kJ/mol)$
  - $2B(s)+3H_2(g)\rightarrow B_2H_6(g)\ (\Delta H = 36\ kJ/mol)$

Find the ΔH_f of:

- $2B(s)+(3/2)O_2(g)\rightarrow B_2O_3(s)$

After the multiplication and reversing of the equations (and their enthalpy changes), the result is:

- $B_2H_6(g)+3O_2(g)\rightarrow B_2O_3(s)+3H_2O(g)\ (\Delta H = -2035\ kJ/mol)$
- $3H_2O(g)\rightarrow 3H_2O(l)\ (\Delta H = -132\ kJ/mol)$
- $3H_2O(l)\rightarrow 3H_2(g)+(3/2)O_2(g)(\Delta H = 858\ kJ/mol)$
- $2B(s)+3H_2(g)\rightarrow B_2H_6(g)(\Delta H = 36\ kJ/mol)$

Adding these equations and canceling out the common terms on both sides, we get

- $2B(s)+(3/2)O_2(g)\rightarrow B_2O_3(s)\ (\Delta H=-1273\ kJ/mol)$
- $C_{graphite}+O_2\rightarrow CO_2(g);(\Delta H=-393.5\ kJ/mol)$ (direct step)
  - $C_{graphite}+1/2\,O_2\rightarrow CO(g);\ (\Delta H=-110.5\ kJ/mol)$
  - $CO(g)+1/2\,O_2\rightarrow CO_2(g);(\Delta H=-283.02\ kJ/mol)$

→In the reactions b) and c), the total ΔH = -393.5 kJ/mol which is equal to ΔH in a)

The difference in the value of ΔH is 0.02 kJ/mol which is due to measurement errors.

Extension to Free Energy and Entropy

The concepts of Hess' law can be expanded to include changes in entropy and in Gibbs free energy, which are also state functions. The Bordwell thermodynamic cycle is an example of such an extension which takes advantage of easily measured equilibria and redox potentials to determine experimentally inaccessible Gibbs free energy values. Combining ΔG° values from Bordwell thermodynamic cycles and ΔH° values found with Hess' law can be helpful in determining entropy values which are not measured directly, and therefore must be calculated through alternative paths.

For the free energy:

$$\Delta G^{\ominus}_{reaction} = \sum \Delta G^{\ominus}_{\mathrm{f}(products)} - \sum \Delta G^{\ominus}_{\mathrm{f}(reactants)}$$

For entropy, the situation is a little different. Because entropy can be measured as an absolute value, not relative to those of the elements in their reference states (as with ΔH° and ΔG°), there is no need to use the entropy of formation; one simply uses the absolute entropies for products and reactants:

$$\Delta S^{\ominus}_{reaction} = \sum S^{\ominus}_{(products)} - \sum S^{\ominus}_{(reactants)}.$$

Applications

Hess' Law of Constant Heat Summation is useful in the determination of enthalpies of the following:

- Heats of formation of unstable intermediates like $CO_{(g)}$ and $NO_{(g)}$.
- Heat changes in phase transitions and allotropic transitions.
- Lattice energies of ionic substances by constructing Born-Haber cycles if the electron affinity to form the anion is known, or
- Electron affinities using a Born-Haber cycle with a theoretical lattice energy

Heat

In thermodynamics, heat is energy in transfer to or from a thermodynamic system, by mechanisms other than thermodynamic work or transfer of matter. The mechanisms include conduction, through direct contact of immobile bodies, or through a wall or barrier that is impermeable to matter; or radiation between separated bodies; or isochoric mechanical work done by the surroundings on the system of interest; or Joule heating by an electric current driven through the system of interest by an external system; or a combination of these. When there is a suitable path between two systems with different temperatures, heat transfer occurs necessarily, immediately, and spontaneously from the hotter to the colder system. Thermal conduction occurs by the stochastic (random) motion of microscopic particles (such as atoms or molecules). In contrast, thermodynamic work is defined by mechanisms that act macroscopically and directly on the system's whole-body state variables; for example, change of the system's volume through a piston's motion with externally measurable force; or change of the system's internal electric polarization through an externally measurable change in electric field. The definition of heat transfer does not require that the process be in any sense smooth. For example, a bolt of lightning may transfer heat to a body.

Convective circulation allows one body to heat another, through an intermediate circulating fluid that carries energy from a boundary of one to a boundary of the other; the actual heat transfer is by conduction and radiation between the fluid and the respective bodies. Natural convection, though spontaneous, does not occur merely because of temperature difference but due to Rayleigh instability caused by combination of gravity and thermal expansion in supplement to sufficient temperature gradient.

Like thermodynamic work, heat transfer is a process involving two systems, not a property of any one system. In thermodynamics, energy transferred as heat (a process function) contributes to change in the system's cardinal energy variable of state, for example its internal energy, or for example its enthalpy. This is to be distinguished from the ordinary language conception of heat as a property of the system.

Although heat flows spontaneously from a hotter body to a cooler one, it is possible to construct a heat pump which expends work to transfer energy from a colder body to a hotter body. In contrast, a heat engine reduces an existing temperature difference to supply work to another system. Another thermodynamic type of heat transfer device is an active heat spreader, which expends work to speed up transfer of energy to colder surroundings from a hotter body, for example a computer component.

The amount of heat transferred in any process can be defined as the total amount of transferred energy excluding any macroscopic work that was done and any energy contained in matter transferred. For the precise definition of heat, it is necessary that it occur by a path that does not include transfer of matter. As an amount of energy (being transferred), the SI unit of heat is the joule (J). The conventional symbol used to represent the amount of heat transferred in a thermodynamic process is Q. Heat is measured by its effect on the states of interacting bodies, for example, by the amount of ice melted or a change in temperature. The quantification of heat via the temperature change of a body is called calorimetry.

The Sun and Earth form an ongoing example of a heating process. Some of the Sun's thermal radiation strikes and heats the Earth. Compared to the Sun, Earth has a much lower temperature and so sends far less thermal radiation back to the Sun. The heat of this process can be quantified by the net amount, and direction (Sun to Earth), of energy it transferred in a given period of time.

Notation and Units

As a form of energy, heat has the unit joule (J) in the International System of Units (SI). However, in many applied fields in engineering the British thermal unit (BTU) and the calorie are often used. The standard unit for the rate of heat transferred is the watt (W), defined as one joule per second.

Use of the symbol Q for the total amount of energy transferred as heat is due to Rudolf Clausius in 1850:

"Let the amount of heat which must be imparted during the transition of the gas in a definite manner from any given state to another, in which its volume is v and its temperature t, be called Q"

Heat released by a system into its surroundings is by convention a negative quantity ($Q < 0$); when a system absorbs heat from its surroundings, it is positive ($Q > 0$). Heat transfer rate, or heat flow per unit time, is denoted by $\dot{Q}$. This should not be confused with a time derivative of a function of state (which can also be written with the dot notation) since heat is not a function of state. Heat flux is defined as rate of heat transfer per unit cross-sectional area (units watts per square metre).

Internal Energy

One of the thermodynamic properties of a system is its internal energy, E, which is the sum of the kinetic and potential energies of the particles that form the system. The internal energy of a system can be understood by examining the simplest possible system: an ideal gas. Because the particles in an ideal gas do not interact, this system has no potential energy. The internal energy of an ideal gas is therefore the sum of the kinetic energies of the particles in the gas.

The kinetic molecular theory assumes that the temperature of a gas is directly proportional to the average kinetic energy of its particles, as shown in the figure below.

$$E_{sys} = {}^{3}/_{2}\, RT$$

In this equation, R is the ideal gas constant in joules per mole kelvin (J/mol-K) and T is the temperature in kelvin.

The internal energy of systems that are more complex than an ideal gas can't be measured directly. But the internal energy of the system is still proportional to its temperature. We can therefore monitor changes in the internal energy of a system by watching what happens to the temperature of the system. Whenever the temperature of the system increases we can conclude that the internal energy of the system has also increased.

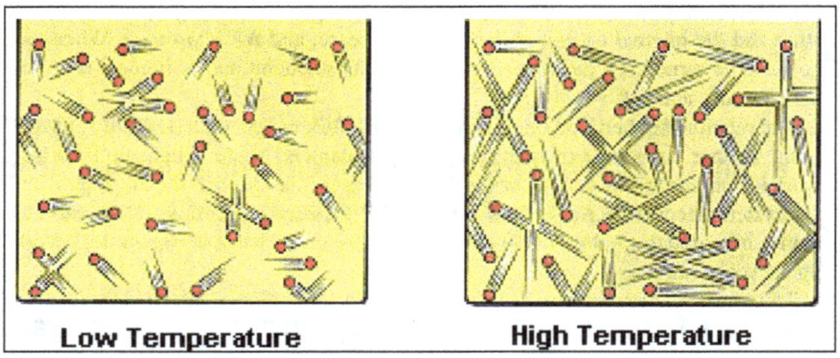

The internal energy of an ideal gas is therefore directly proportional to the temperature of the gas.

Assume, for the moment, that a thermometer immersed in a beaker of water on a hot plate reads 73.5°C, as shown in the figure below. This measurement can only describe the state of the system at that moment in time. It can't tell us whether the water was heated directly from room temperature to 73.5°C or heated from room temperature to 100°C and then allowed to cool.

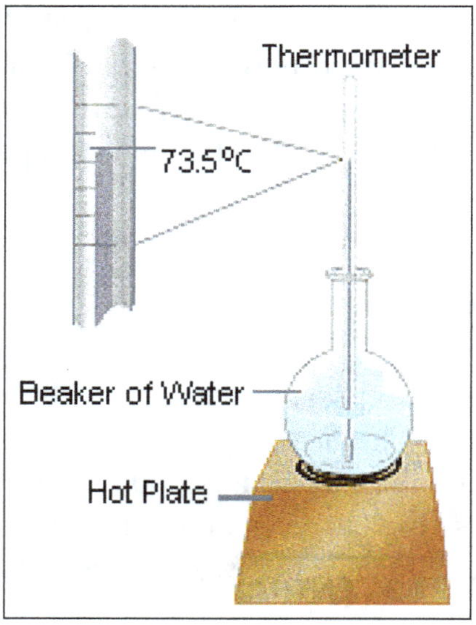

Temperature is therefore a state function. It depends only on the state of the system at any moment in time, not the path used to get the system to that state. Because the internal energy of the system is proportional to its temperature, internal energy is also a state function. Any change in the internal energy of the system is equal to the difference between its initial and final values.

$$\Delta E_{sys} = E_f - E_i$$

Enthalpy

Enthalpy a property of a thermodynamic system, is equal to the system's internal energy plus the product of its pressure and volume. In a system enclosed so as to prevent mass transfer, for processes at constant pressure, the heat absorbed or released equals the change in enthalpy.

The unit of measurement for enthalpy in the International System of Units (SI) is the joule. Other historical conventional units still in use include the British thermal unit (BTU) and the calorie.

Enthalpy comprises a system's internal energy, which is the energy required to create the system, plus the amount of work required to make room for it by displacing its environment and establishing its volume and pressure.

Enthalpy is defined as a state function that depends only on the prevailing equilibrium state identified by the system's internal energy, pressure, and volume. It is an extensive quantity.

Change in enthalpy (ΔH) is the preferred expression of system energy change in many chemical, biological, and physical measurements at constant pressure, because it simplifies the description of energy transfer. In a system enclosed so as to prevent matter transfer, at constant pressure, the enthalpy change equals the energy transferred from the environment through heat transfer or work other than expansion work.

The total enthalpy, H, of a system cannot be measured directly. The same situation exists in classical mechanics: only a change or difference in energy carries physical meaning. Enthalpy itself is a thermodynamic potential, so in order to measure the enthalpy of a system, we must refer to a defined reference point; therefore what we measure is the change in enthalpy, ΔH. The ΔH is a positive change in endothermic reactions, and negative in heat-releasing exothermic processes.

For processes under constant pressure, ΔH is equal to the change in the internal energy of the system, plus the pressure-volume work $p\,\Delta V$ done by the system on its surroundings (which is > 0 for an expansion and < 0 for a contraction). This means that the change in enthalpy under such conditions is the heat absorbed or released by the system through a chemical reaction or by external heat transfer. Enthalpies for chemical substances at constant pressure usually refer to standard state: most commonly 1 bar pressure. Standard state does not, strictly speaking, specify a temperature, but expressions for enthalpy generally reference the standard heat of formation at 25 °C.

Enthalpy of ideal gases and incompressible solids and liquids does not depend on pressure, unlike entropy and Gibbs energy. Real materials at common temperatures and pressures usually closely approximate this behavior, which greatly simplifies enthalpy calculation and use in practical designs and analyses.

The enthalpy of a thermodynamic system is defined as:

$$H = U + pV,$$

where,

H is enthalpy,

U is the internal energy of the system,

p is pressure,

V is the volume of the system.

Enthalpy is an extensive property. This means that, for homogeneous systems, the enthalpy is proportional to the size of the system. It is convenient to introduce the specific enthalpy $h = \frac{H}{m}$, where m is the mass of the system, or the molar enthalpy $H_{\mathrm{m}} = \frac{H}{n}$, where n is the number of moles (h and H_{m} are intensive properties). For inhomogeneous systems the enthalpy is the sum of the enthalpies of the composing subsystems:

$$H = \sum_k H_k,$$

where:

H is the total enthalpy of all the subsystems,

k refers to the various subsystems,

H_k refers to the enthalpy of each subsystem,

$\sum_k H_k$ refers to the sum of the enthalpies of all subsystems.

A closed system may lie in thermodynamic equilibrium in a static gravitational field, so that its pressure p varies continuously with altitude, while, because of the equilibrium requirement, its temperature T is invariant with altitude. (Correspondingly, the system's gravitational potential energy density also varies with altitude.) Then the enthalpy summation becomes an integral:

$$H = \int (\rho h) dV$$

where:

p ("rho") is density (mass per unit volume),

h is the specific enthalpy (enthalpy per unit mass),

(ph) represents the enthalpy density (enthalpy per unit volume),

dV denotes an infinitesimally small element of volume within the system, for example, the volume of an infinitesimally thin horizontal layer,

$\int(\rho h)dV$ represents the sum of the enthalpies of all the elements of the volume.

The enthalpy of a closed homogeneous system is its cardinal energy function $H(S,p)$, with natural state variables its entropy $S[p]$ and its pressure p. A differential relation for it can be derived as follows. We start from the first law of thermodynamics for closed systems for an infinitesimal process:

$$dU = \delta Q - \delta W \text{ ,}$$

where:

δQ is a small amount of heat added to the system,

δW a small amount of work performed by the system.

In a homogeneous system in which only reversible, or quasi-static, processes are considered, the second law of thermodynamics gives $\delta Q = T\,dS$, with T the absolute temperature of the system. Furthermore, if only pV work is done, $\delta W = pdV$. As a result,

$$dU = T\,dS - pdV \text{ .}$$

Adding $d(pV)$ to both sides of this expression gives,

$$dU + d(pV) = T\,dS - pdV + d(pV),$$

or

$$d(U + pV) = T\,dS + V\,dp,$$

so

$$dH(S,p) = T\,dS + V\,dp.$$

Other Expressions

The above expression of dH in terms of entropy and pressure may be unfamiliar to some readers. However, there are expressions in terms of more familiar variables such as temperature and pressure:

$$dH = C_p\,dT + V(1 - \alpha T)dp.$$

Here C_p is the heat capacity at constant pressure and α is the coefficient of (cubic) thermal expansion:

$$\alpha = \frac{1}{V}\left(\frac{\partial V}{\partial T}\right)_p .$$

With this expression one can, in principle, determine the enthalpy if C_p and V are known as functions of p and T.

Note that for an ideal gas, $\alpha T = 1$, so that,

$$dH = C_p dT.$$

In a more general form, the first law describes the internal energy with additional terms involving the chemical potential and the number of particles of various types. The differential statement for dH then becomes,

$$dH = T\,dS + V\,dp + \sum_i \mu_i dN_i,$$

where μ_i is the chemical potential per particle for an i-type particle, and N_i is the number of such particles. The last term can also be written as $\mu_i dn_i$ (with dn_i the number of moles of component i added to the system and, in this case, μ_i the molar chemical potential) or as $\mu_i dm_i$ (with dm_i the mass of component i added to the system and, in this case, μ_i the specific chemical potential).

Cardinal Functions

The enthalpy, $H(S[p],p,\{N_i\})$, expresses the thermodynamics of a system in the *energy-language*, or in the *energy representation*. As a function of state, its arguments include both one intensive and several extensive state variables. The state variables $S[p]$, p, and $\{N_i\}$ are said to be the *natural state variables* in this representation. They are suitable for describing processes in which they are experimentally controlled. For example, in an idealized process, $S[p]$ and p can be controlled by preventing heat and matter transfer by enclosing the system with a wall that is adiathermal and impermeable to matter, and by making the process infinitely slow, and by varying only the external pressure on the piston that controls the volume of the system. This is the basis of the so-called adiabatic approximation that is used in meteorology.

Alongside the enthalpy, with these arguments, the other cardinal function of state of a thermodynamic system is its entropy, as a function, $S[p](H,p,\{N_i\})$, of the same list of variables of state, except that the entropy, $S[p]$, is replaced in the list by the enthalpy, H. It expresses the *entropy representation*. The state variables H, p, and $\{N_i\}$ are said to be the *natural state variables* in this representation. They are suitable for describing processes in which they are experimentally controlled. For example, H and p can be controlled by allowing heat transfer, and by varying only the external pressure on the piston that sets the volume of the system.

Physical Interpretation

The U term can be interpreted as the energy required to create the system, and the pV term as the work that would be required to "make room" for the system if the pressure of the environment remained constant. When a system, for example, n moles of a gas of volume V at pressure p and temperature T, is created or brought to its present state from absolute zero, energy must be supplied equal to its internal energy U plus pV, where pV is the work done in pushing against the ambient (atmospheric) pressure.

In basic physics and statistical mechanics it may be more interesting to study the internal properties of the system and therefore the internal energy is used. In basic chemistry, experiments are often conducted at constant atmospheric pressure, and the pressure-volume work represents an energy exchange with the atmosphere that cannot be accessed or controlled, so that ΔH is the expression chosen for the heat of reaction.

For a heat engine a change in its internal energy is the difference between the heat input and the pressure-volume work done by the working substance while a change in its enthalpy is the difference between the heat input and the work done by the engine:

$$dH = \delta Q - \delta W$$

where the work W done by the engine is:

$$W = -\oint pdV$$

Relationship to Heat

In order to discuss the relation between the enthalpy increase and heat supply, we return to the first law for closed systems, with the physics sign convention: $dU = \delta Q - \delta W$, where the heat δQ is supplied by conduction, radiation, and Joule heating. We apply it to the special case with a constant pressure at the surface. In this case the work term can be split into two contributions, the so-called pV work, given by $p\,dV$ (where here p is the pressure at the surface, dV is the increase of the volume of the system), and the so-called isochoric mechanical work $\delta W'$, such as stirring by a shaft with paddles or by an externally driven magnetic field acting on an internal rotor. Cases of long range electromagnetic interaction require further state variables in their formulation, and are not considered here. So we write $\delta W = p\,dV + \delta W'$. In this case the first law reads:

$$dU = \delta Q - pdV - \delta W'.$$

Now,

$$dH = dU + d(pV).$$

So

$$dH = \delta Q + V\,dp + p dV - p dV - \delta W'$$

$$= \delta Q + V\,dp - \delta W'.$$

With sign convention of physics, $\delta W' < 0$, because isochoric shaft work done by an external device on the system adds energy to the system, and may be viewed as virtually adding heat. The only thermodynamic mechanical work done by the system is expansion work, $p\,dV$.

The system is under constant pressure $(dp = 0)$. Consequently, the increase in enthalpy of the system is equal to the added heat and virtual heat:

$$dH = \delta Q - \delta W'.$$

This is why the now obsolete term *heat content* was used in the nineteenth century.

Applications

In thermodynamics, one can calculate enthalpy by determining the requirements for creating a system from "nothingness"; the mechanical work required, pV, differs based upon the conditions that obtain during the creation of the thermodynamic system.

Energy must be supplied to remove particles from the surroundings to make space for the creation of the system, assuming that the pressure p remains constant; this is the pV term. The supplied energy must also provide the change in internal energy, U, which includes activation energies, ionization energies, mixing energies, vaporization energies, chemical bond energies, and so forth. Together, these constitute the change in the enthalpy $U + pV$. For systems at constant pressure, with no external work done other than the pV work, the change in enthalpy is the heat received by the system.

For a simple system, with a constant number of particles, the difference in enthalpy is the maximum amount of thermal energy derivable from a thermodynamic process in which the pressure is held constant.

Heat of Reaction

The total enthalpy of a system cannot be measured directly, the *enthalpy change* of a system is measured instead. Enthalpy change is defined by the following equation:

$$\Delta H = H_\mathrm{f} - H_\mathrm{i},$$

where:

ΔH is the "enthalpy change",

H_f is the final enthalpy of the system (in a chemical reaction, the enthalpy of the products),

H_i is the initial enthalpy of the system (in a chemical reaction, the enthalpy of the reactants).

For an exothermic reaction at constant pressure, the system's change in enthalpy equals the energy released in the reaction, including the energy retained in the system and lost through expansion against its surroundings. In a similar manner, for an endothermic reaction, the system's change in enthalpy is equal to the energy *absorbed* in the reaction, including the energy *lost by* the system and *gained* from compression from its surroundings. If ΔH is positive, the reaction is endothermic, that is heat is absorbed by the system due to the products of the reaction having a greater enthalpy than the reactants. On the other hand, if ΔH is negative, the reaction is exothermic, that is the overall decrease in enthalpy is achieved by the generation of heat.

From the definition of enthalpy as $H = U + pV$ the enthalpy change at constant pressure $\Delta H = \Delta U + p\Delta V$. However for most chemical reactions, the work term $p\Delta V$ is much smaller than the internal energy change ΔU which is approximately equal to ΔH As an example, for the combustion of carbon monoxide 2 $CO(g)$ + $O_2(g)$ → 2 $CO_2(g)$, $\Delta H = -566.0$ kJ and $\Delta U = -563.5$ kJ. Since the differences are so small, reaction enthalpies are often loosely described as reaction energies and analyzed in terms of bond energies.

Specific Enthalpy

The specific enthalpy of a uniform system is defined as $h = \frac{H}{m}$ where m is the mass of the system. The SI unit for specific enthalpy is joule per kilogram. It can be expressed in other specific quantities by $h = u + pv$, where u is the specific internal energy, p is the pressure, and v is specific volume, which is equal to $\frac{1}{\rho}$, where ρ is the density.

Enthalpy Changes

An enthalpy change describes the change in enthalpy observed in the constituents of a thermodynamic system when undergoing a transformation or chemical reaction. It is the difference between the enthalpy after the process has completed, i.e. the enthalpy of the products, and the initial enthalpy of the system, i.e. the reactants. These processes are reversible and the enthalpy for the reverse process is the negative value of the forward change.

A common standard enthalpy change is the enthalpy of formation, which has been determined for a large number of substances. Enthalpy changes are routinely measured and compiled in chemical and physical reference works, such as the CRC Handbook of Chemistry and Physics. The following is a selection of enthalpy changes commonly recognized in thermodynamics.

When used in these recognized terms the qualifier *change* is usually dropped and the property is simply termed *enthalpy of 'process'*. Since these properties are often used as reference values it is very common to quote them for a standardized set of environmental parameters, or standard conditions, including:

- A temperature of 25 °C or 298.15 K,
- A pressure of one atmosphere (1 atm or 101.325 kPa),
- A concentration of 1.0 M when the element or compound is present in solution,
- Elements or compounds in their normal physical states, i.e. standard state.

For such standardized values the name of the enthalpy is commonly prefixed with the term *standard*, e.g. *standard enthalpy of formation*.

Chemical properties:

- Enthalpy of reaction, defined as the enthalpy change observed in a constituent of a thermodynamic system when one mole of substance reacts completely.
- Enthalpy of formation, defined as the enthalpy change observed in a constituent of a thermodynamic system when one mole of a compound is formed from its elementary antecedents.
- Enthalpy of combustion, defined as the enthalpy change observed in a constituent of a thermodynamic system when one mole of a substance burns completely with oxygen.
- Enthalpy of hydrogenation, defined as the enthalpy change observed in a constituent of a thermodynamic system when one mole of an unsaturated compound reacts completely with an excess of hydrogen to form a saturated compound.
- Enthalpy of atomization, defined as the enthalpy change required to atomize one mole of compound completely.
- Enthalpy of neutralization, defined as the enthalpy change observed in a constituent of a thermodynamic system when one mole of water is formed when an acid and a base react.
- Standard Enthalpy of solution, defined as the enthalpy change observed in a constituent of a thermodynamic system when one mole of a solute is dissolved completely in an excess of solvent, so that the solution is at infinite dilution.
- Standard enthalpy of Denaturation (biochemistry), defined as the enthalpy change required to denature one mole of compound.

- Enthalpy of hydration, defined as the enthalpy change observed when one mole of gaseous ions are completely dissolved in water forming one mole of aqueous ions.

Physical properties:

- Enthalpy of fusion, defined as the enthalpy change required to completely change the state of one mole of substance between solid and liquid states.
- Enthalpy of vaporization, defined as the enthalpy change required to completely change the state of one mole of substance between liquid and gaseous states.
- Enthalpy of sublimation, defined as the enthalpy change required to completely change the state of one mole of substance between solid and gaseous states.
- Lattice enthalpy, defined as the energy required to separate one mole of an ionic compound into separated gaseous ions to an infinite distance apart (meaning no force of attraction).
- Enthalpy of mixing, defined as the enthalpy change upon mixing of two (non-reacting) chemical substances.

Open Systems

In thermodynamic open systems, mass (of substances) may flow in and out of the system boundaries. The first law of thermodynamics for open systems states: The increase in the internal energy of a system is equal to the amount of energy added to the system by mass flowing in and by heating, minus the amount lost by mass flowing out and in the form of work done by the system:

$$dU = \delta Q + dU_{\text{in}} - dU_{\text{out}} - \delta W ,$$

where U_{in} is the average internal energy entering the system, and U_{out} is the average internal energy leaving the system.

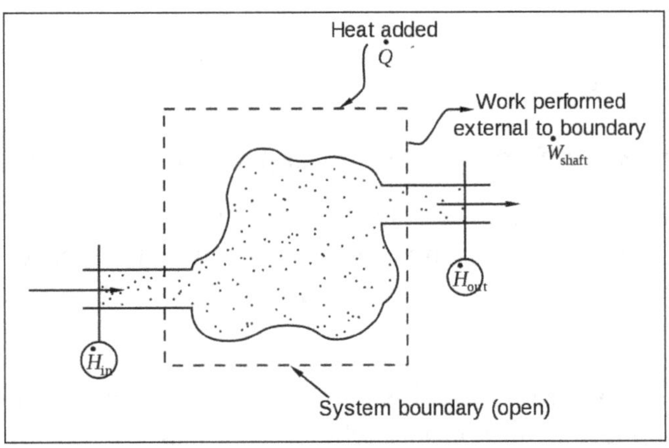

During steady, continuous operation, an energy balance applied to an open system equates shaft work performed by the system to heat added plus net enthalpy added

The region of space enclosed by the boundaries of the open system is usually called a control volume, and it may or may not correspond to physical walls. If we choose the shape of the control volume such that all flow in or out occurs perpendicular to its surface, then the flow of mass into the system performs work as if it were a piston of fluid pushing mass into the system, and the system performs work on the flow of mass out as if it were driving a piston of fluid. There are then two types of work performed: *flow work* described above, which is performed on the fluid (this is also often called *pV work*), and *shaft work*, which may be performed on some mechanical device.

These two types of work are expressed in the equation:

$$\delta W = d(p_{\text{out}}V_{\text{out}}) - d(p_{\text{in}}V_{\text{in}}) + \delta W_{\text{shaft}}.$$

Substitution into the equation above for the control volume (cv) yields:

$$dU_{\text{cv}} = \delta Q + dU_{\text{in}} + d(p_{\text{in}}V_{\text{in}}) - dU_{\text{out}} - d(p_{\text{out}}V_{\text{out}}) - \delta W_{\text{shaft}}.$$

The definition of enthalpy, H, permits us to use this thermodynamic potential to account for both internal energy and pV work in fluids for open systems:

$$dU_{\text{cv}} = \delta Q + dH_{\text{in}} - dH_{\text{out}} - \delta W_{\text{shaft}}.$$

If we allow also the system boundary to move (e.g. due to moving pistons), we get a rather general form of the first law for open systems. In terms of time derivatives it reads:

$$\frac{dU}{dt} = \sum_k \dot{Q}_k + \sum_k \dot{H}_k - \sum_k p_k \frac{dV_k}{dt} - P,$$

with sums over the various places k where heat is supplied, mass flows into the system, and boundaries are moving. The $\dot{H}_k$ terms represent enthalpy flows, which can be written as,

$$\dot{H}_k = h_k \dot{m}_k = H_{\text{m}} \dot{n}_k,$$

with $\dot{m}_k$ the mass flow and $\dot{n}_k$ the molar flow at position k respectively. The term $\frac{dV_k}{dt}$ represents the rate of change of the system volume at position k that results in pV power done by the system. The parameter P represents all other forms of power done by the system such as shaft power, but it can also be e.g. electric power produced by an electrical power plant.

Note that the previous expression holds true only if the kinetic energy flow rate is conserved between system inlet and outlet. Otherwise, it has to be included in the enthalpy balance. During steady-state operation of a device, the average $\frac{dU}{dt}$ may be set equal to zero. This yields a useful expression for the average power generation for these devices in the absence of chemical reactions:

$$P = \sum_k \langle Q_k \rangle + \sum_k \langle H_k \rangle - \sum_k \left\langle p_k \frac{dV}{dt} \right\rangle$$

where the angle brackets denote time averages. The technical importance of the enthalpy is directly related to its presence in the first law for open systems, as formulated above.

Entropy

In statistical mechanics, entropy is an extensive property of a thermodynamic system. It is closely related to the number Ω of microscopic configurations (known as microstates) that are consistent with the macroscopic quantities that characterize the system (such as its volume, pressure and temperature). Entropy expresses the number Ω of different configurations that a system defined by macroscopic variables could assume. Under the assumption that each microstate is equally probable, the entropy is the natural logarithm of the number of microstates, multiplied by the Boltzmann constant k_B. Formally (assuming equiprobable microstates),

$$S = k_B \ln \Omega.$$

Macroscopic systems typically have a very large number Ω of possible microscopic configurations. For example, the entropy of an ideal gas is proportional to the number of gas molecules *N*. The number of molecules in twenty liters of gas at room temperature and atmospheric pressure is roughly $N \approx 6 \times 10^{23}$ (the Avogadro number).

The second law of thermodynamics states that the entropy of an isolated system never decreases over time. Isolated systems spontaneously evolve towards thermodynamic equilibrium, the state with maximum entropy. Non-isolated systems, like organisms, may lose entropy, provided their environment's entropy increases by at least that amount so that the total entropy increases. Therefore, total entropy in the Universe does increase. Entropy is a function of the state of the system, so the change in entropy of a system is determined by its initial and final states. In the idealization that a process is reversible, the entropy does not change, while irreversible processes always increase the total entropy.

Because it is determined by the number of random microstates, entropy is related to the amount of additional information needed to specify the exact physical state of a system, given its macroscopic specification. For this reason, it is often said that entropy is an expression of the disorder, or randomness of a system, or of the lack of information about it.

Applications

The Fundamental Thermodynamic Relation

The entropy of a system depends on its internal energy and its external parameters, such as its volume. In the thermodynamic limit, this fact leads to an equation relating the change in the internal energy U to changes in the entropy and the external parameters. This relation is known as the *fundamental thermodynamic relation*. If external pressure p bears on the volume V as the only external parameter, this relation is:

$$dU = T\,dS - p\,dV$$

Since both internal energy and entropy are monotonic functions of temperature T, implying that the internal energy is fixed when one specifies the entropy and the volume, this relation is valid even if the change from one state of thermal equilibrium to another with infinitesimally larger entropy and volume happens in a non-quasistatic way (so during this change the system may be very far out of thermal equilibrium and then the entropy, pressure and temperature may not exist).

The fundamental thermodynamic relation implies many thermodynamic identities that are valid in general, independent of the microscopic details of the system. Important examples are the Maxwell relations and the relations between heat capacities.

Entropy in Chemical Thermodynamics

Thermodynamic entropy is central in chemical thermodynamics, enabling changes to be quantified and the outcome of reactions predicted. The second law of thermodynamics states that entropy in an isolated system – the combination of a subsystem under study and its surroundings – increases during all spontaneous chemical and physical processes. The Clausius equation of $\delta q_{rev} / T = \Delta S$ introduces the measurement of entropy change, ΔS. Entropy change describes the direction and quantifies the magnitude of simple changes such as heat transfer between systems – always from hotter to cooler spontaneously.

The thermodynamic entropy therefore has the dimension of energy divided by temperature, and the unit joule per kelvin (J/K) in the International System of Units (SI).

Thermodynamic entropy is an extensive property, meaning that it scales with the size or extent of a system. In many processes it is useful to specify the entropy as an intensive property independent of the size, as a specific entropy characteristic of the type of system studied. Specific entropy may be expressed relative to a unit of mass, typically the kilogram (unit: $J{\cdot}kg^{-1}{\cdot}K^{-1}$). Alternatively, in chemistry, it is also referred to one mole of substance, in which case it is called the *molar entropy* with a unit of $J{\cdot}mol^{-1}{\cdot}K^{-1}$.

Thus, when one mole of substance at about 0 K is warmed by its surroundings to 298 K, the sum of the incremental values of q_{rev}/T constitute each element's or compound's standard molar entropy, an indicator of the amount of energy stored by a substance at 298 K. Entropy change also measures the mixing of substances as a summation of their relative quantities in the final mixture.

Entropy is equally essential in predicting the extent and direction of complex chemical reactions. For such applications, ΔS must be incorporated in an expression that includes both the system and its surroundings, $\Delta S_{universe} = \Delta S_{surroundings} + \Delta S_{system}$. This expression becomes, via some steps, the Gibbs free energy equation for reactants and products in the system: ΔG [the Gibbs free energy change of the system] = ΔH [the enthalpy change] – $T\Delta S$ [the entropy change].

Entropy Balance Equation for Open Systems

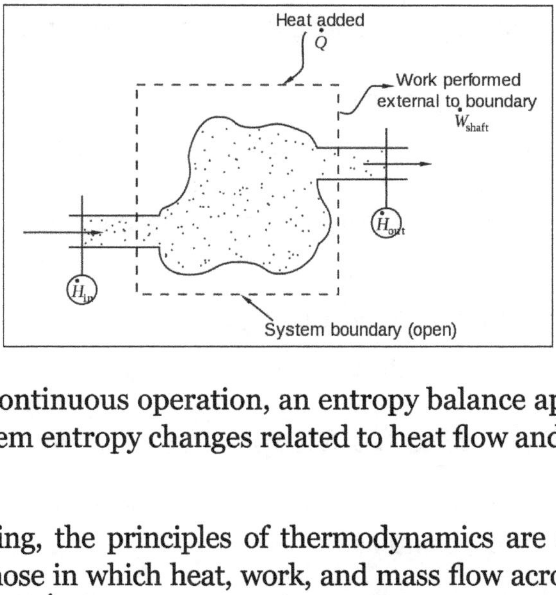

During steady-state continuous operation, an entropy balance applied to an open system accounts for system entropy changes related to heat flow and mass flow across the system boundary.

In chemical engineering, the principles of thermodynamics are commonly applied to "open systems", i.e. those in which heat, work, and mass flow across the system boundary. Flows of both heat ($\dot{Q}$) and work, i.e. $\dot{W}_S$ (shaft work) and $P(dV/dt)$ (pressure-volume work), across the system boundaries, in general cause changes in the entropy of the system. Transfer as heat entails entropy transfer $\dot{Q}/T$ where T is the absolute thermodynamic temperature of the system at the point of the heat flow. If there are mass flows across the system boundaries, they also influence the total entropy of the system. This account, in terms of heat and work, is valid only for cases in which the work and heat transfers are by paths physically distinct from the paths of entry and exit of matter from the system.

To derive a generalized entropy balanced equation, we start with the general balance equation for the change in any extensive quantity Θ in a thermodynamic system, a quantity that may be either conserved, such as energy, or non-conserved, such as entropy. The basic generic balance expression states that $d\Theta/dt$, i.e. the rate of change of

Θ in the system, equals the rate at which Θ enters the system at the boundaries, minus the rate at which Θ leaves the system across the system boundaries, plus the rate at which Θ is generated within the system. For an open thermodynamic system in which heat and work are transferred by paths separate from the paths for transfer of matter, using this generic balance equation, with respect to the rate of change with time *t* of the extensive quantity entropy *S*, the entropy balance equation is:

$$\frac{dS}{dt} = \sum_{k=1}^{K} \dot{M}_k \hat{S}_k + \frac{\dot{Q}}{T} + \dot{S}_{\text{gen}}$$

where:

$\sum_{k=1}^{K} \dot{M}_k \hat{S}_k = t$ the net rate of entropy flow due to the flows of mass into and out of the system (where $\hat{S}$ = entropy per unit mass).

$\frac{\dot{Q}}{T}$ = the rate of entropy flow due to the flow of heat across the system boundary.

$\dot{S}_{\text{gen}}$ = the rate of entropy production within the system. This entropy production arises from processes within the system, including chemical reactions, internal matter diffusion, internal heat transfer, and frictional effects such as viscosity occurring within the system from mechanical work transfer to or from the system.

Note, also, that if there are multiple heat flows, the term $\dot{Q}/T$ is replaced by $\sum \dot{Q}_j / T_j$ where $\dot{Q}_j$ is the heat flow and T_j is the temperature at the *j*th heat flow port into the system.

Entropy Change Formulas for Simple Processes

For certain simple transformations in systems of constant composition, the entropy changes are given by simple formulas.

Isothermal Expansion or Compression of an Ideal Gas

For the expansion (or compression) of an ideal gas from an initial volume V_0 and pressure P_0 to a final volume *V* and pressure *P* at any constant temperature, the change in entropy is given by:

$$\Delta S = nR \ln \frac{V}{V_0} = -nR \ln \frac{P}{P_0}.$$

Here *n* is the number of moles of gas and *R* is the ideal gas constant. These equations also apply for expansion into a finite vacuum or a throttling process, where the temperature, internal energy and enthalpy for an ideal gas remain constant.

Cooling and Heating

For heating or cooling of any system (gas, liquid or solid) at constant pressure from an initial temperature T_0 to a final temperature T, the entropy change is:

$$\Delta S = nC_P \ln \frac{T}{T_0}.$$

provided that the constant-pressure molar heat capacity (or specific heat) C_P is constant and that no phase transition occurs in this temperature interval.

Similarly at constant volume, the entropy change is:

$$\Delta S = nC_v \ln \frac{T}{T_0},$$

where the constant-volume molar heat capacity C_v is constant and there is no phase change.

At low temperatures near absolute zero, heat capacities of solids quickly drop off to near zero, so the assumption of constant heat capacity does not apply.

Since entropy is a state function, the entropy change of any process in which temperature and volume both vary is the same as for a path divided into two steps – heating at constant volume and expansion at constant temperature. For an ideal gas, the total entropy change is

$$\Delta S = nC_v \ln \frac{T}{T_0} + nR \ln \frac{V}{V_0}.$$

Similarly if the temperature and pressure of an ideal gas both vary,

$$\Delta S = nC_P \ln \frac{T}{T_0} - nR \ln \frac{P}{P_0}.$$

Phase Transitions

Reversible phase transitions occur at constant temperature and pressure. The reversible heat is the enthalpy change for the transition, and the entropy change is the enthalpy change divided by the thermodynamic temperature. For fusion (melting) of a solid to a liquid at the melting point T_m, the entropy of fusion is,

$$\Delta S_{fus} = \frac{\Delta H_{fus}}{T_m}.$$

Similarly, for vaporization of a liquid to a gas at the boiling point T_b, the entropy of vaporization is,

$$\Delta S_{vap} = \frac{\Delta H_{vap}}{T_b}.$$

Gibbs Free Energy

In thermodynamics, the Gibbs free energy is a thermodynamic potential that can be used to calculate the maximum of reversible work that may be performed by a thermodynamic system at a constant temperature and pressure (isothermal, isobaric). The Gibbs free energy ($\Delta G^{\circ} = \Delta H^{\circ} - T\Delta S^{\circ}$; J in SI units) is the *maximum* amount of non-expansion work that can be extracted from a thermodynamically closed system (one that can exchange heat and work with its surroundings, but not matter); this maximum can be attained only in a completely reversible process. When a system transforms reversibly from an initial state to a final state, the decrease in Gibbs free energy equals the work done by the system to its surroundings, minus the work of the pressure forces.

The Gibbs energy (also referred to as *G*) is also the thermodynamic potential that is minimized when a system reaches chemical equilibrium at constant pressure and temperature. Its derivative with respect to the reaction coordinate of the system vanishes at the equilibrium point. As such, a reduction in *G* is a necessary condition for the spontaneity of processes at constant pressure and temperature.

The Gibbs free energy, originally called *available energy*, was developed in the 1870s by the American scientist Josiah Willard Gibbs. In 1873, Gibbs described this "available energy" as,

> the greatest amount of mechanical work which can be obtained from a given quantity of a certain substance in a given initial state, without increasing its total volume or allowing heat to pass to or from external bodies, except such as at the close of the processes are left in their initial condition.

The initial state of the body, according to Gibbs, is supposed to be such that "the body can be made to pass from it to states of dissipated energy by reversible processes".

The reaction $C_{(s)}^{diamond} \rightarrow C_{(s)}^{graphite}$ has a negative change in Gibbs free energy and is therefore thermodynamically favorable at 25 °C and 1 atm. However, even though favorable, it is so slow that it is not observed. Whether a reaction is thermodynamically favorable does not determine its rate.

According to the second law of thermodynamics, for systems reacting at STP (or any other fixed temperature and pressure), there is a general natural tendency to achieve a minimum of the Gibbs free energy.

A quantitative measure of the favorability of a given reaction at constant temperature and pressure is the change ΔG (sometimes written "delta G" or "dG") in Gibbs free energy that is (or would be) caused by the reaction. As a necessary condition for the reaction to occur at constant temperature and pressure, ΔG must be smaller than the non-PV (e.g. electrical) work, which is often equal to zero (hence ΔG must be negative). ΔG equals the maximum amount of non-*PV* work that can be performed as a result of the chemical reaction for the case of reversible process. If the analysis indicated a positive ΔG for the reaction, then energy — in the form of electrical or other non-*PV* work — would have to be added to the reacting system for ΔG to be smaller than the non-*PV* work and make it possible for the reaction to occur.

We can think of ΔG as the amount of "free" or "useful" energy available to do work. The equation can be also seen from the perspective of the system taken together with its surroundings (the rest of the universe). First, assume that the given reaction at constant temperature and pressure is the only one that is occurring. Then the entropy released or absorbed by the system equals the entropy that the environment must absorb or release, respectively. The reaction will only be allowed if the total entropy change of the universe is zero or positive. This is reflected in a negative ΔG, and the reaction is called exergonic.

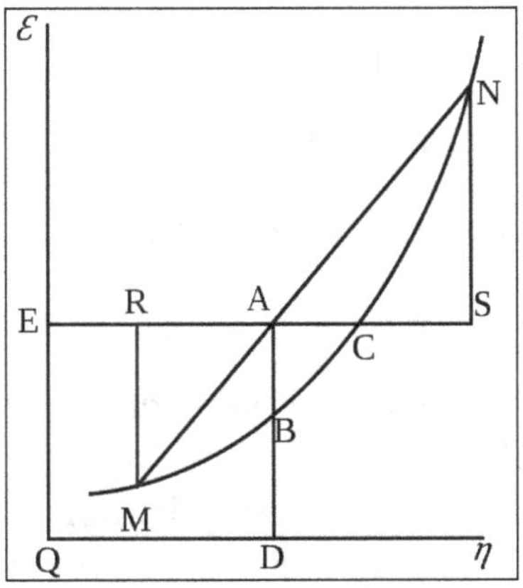

If we couple reactions, then an otherwise endergonic chemical reaction (one with positive ΔG) can be made to happen. The input of heat into an inherently endergonic reaction, such as the elimination of cyclohexanol to cyclohexene, can be seen as coupling an unfavourable reaction (elimination) to a favourable one (burning of coal or other provision of heat) such that the total entropy change of the universe is greater than or equal to zero, making the *total* Gibbs free energy difference of the coupled reactions negative.

In traditional use, the term "free" was included in "Gibbs free energy" to mean "available in the form of useful work". The characterization becomes more precise if we add the qualification that it is the energy available for *non-volume* work. (An analogous, but slightly different, meaning of "free" applies in conjunction with the Helmholtz free energy, for systems at constant temperature).

Willard Gibbs' 1873 available energy (free energy) graph, which shows a plane perpendicular to the axis of v (volume) and passing through point A, which represents the initial state of the body. MN is the section of the surface of dissipated energy. Qε and Qη are sections of the planes $\eta = 0$ and $\varepsilon = 0$, and therefore parallel to the axes of ε (internal energy) and η (entropy), respectively. AD and AE are the energy and entropy of the body in its initial state, AB and AC its *available energy* (Gibbs free energy) and its *capacity for entropy* (the amount by which the entropy of the body can be increased without changing the energy of the body or increasing its volume) respectively.

The Gibbs free energy is defined as

$$G(p,T) = U + pV - TS,$$

which is the same as

$$G(p,T) = H - TS,$$

where:

U is the internal energy (SI unit: joule),

p is pressure (SI unit: pascal),

V is volume (SI unit: m^3),

T is the temperature (SI unit: kelvin),

S is the entropy (SI unit: joule per kelvin),

H is the enthalpy (SI unit: joule).

The expression for the infinitesimal reversible change in the Gibbs free energy as a function of its "natural variables" p and T, for an open system, subjected to the operation of external forces (for instance, electrical or magnetic) X_i, which cause the external parameters of the system a_i to change by an amount $\mathrm{d}a_i$, can be derived as follows from the first law for reversible processes:

$$T\mathrm{d}S = \mathrm{d}U + p\mathrm{d}V - \sum_{i=1}^{k} \mu_i \mathrm{d}N_i + \sum_{i=1}^{n} X_i \mathrm{d}a_i + \cdots$$

$$\mathrm{d}(TS) - S\mathrm{d}T = \mathrm{d}U + \mathrm{d}(pV) - V\mathrm{d}p - \sum_{i=1}^{k} \mu_i \mathrm{d}N_i + \sum_{i=1}^{n} X_i \mathrm{d}a_i + \cdots$$

$$d(U - TS + pV) = V\,dp - S\,dT + \sum_{i=1}^{k} \mu_i\,dN_i - \sum_{i=1}^{n} X_i\,da_i + \cdots$$

$$dG = V\,dp - S\,dT + \sum_{i=1}^{k} \mu_i\,dN_i - \sum_{i=1}^{n} X_i\,da_i + \cdots$$

where:

μ_i is the chemical potential of the *i*th chemical component. (SI unit: joules per particle or joules per mole).

N_i is the number of particles (or number of moles) composing the *i*th chemical component.

This is one form of Gibbs fundamental equation. In the infinitesimal expression, the term involving the chemical potential accounts for changes in Gibbs free energy resulting from an influx or outflux of particles. In other words, it holds for an open system or for a closed, chemically reacting system where the N_i are changing. For a closed, non-reacting system, this term may be dropped.

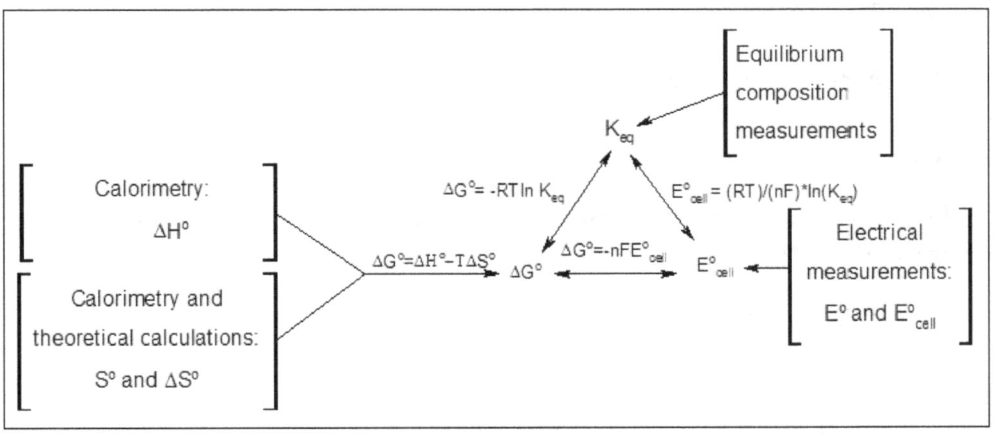

Relation to other relevant parameters.

Any number of extra terms may be added, depending on the particular system being considered. Aside from mechanical work, a system may, in addition, perform numerous other types of work. For example, in the infinitesimal expression, the contractile work energy associated with a thermodynamic system that is a contractile fiber that shortens by an amount $-dl$ under a force f would result in a term $f\,dl$ being added. If a quantity of charge $-de$ is acquired by a system at an electrical potential Ψ, the electrical work associated with this is $-\Psi\,de$, which would be included in the infinitesimal expression. Other work terms are added on per system requirements.

Each quantity in the equations above can be divided by the amount of substance, measured in moles, to form *molar Gibbs free energy*. The Gibbs free energy is one of the most important thermodynamic functions for the characterization of a system. It is a factor in determining outcomes such as the voltage of an electrochemical cell, and the equilibrium constant for a reversible reaction. In isothermal, isobaric systems, Gibbs free energy can be thought of as a "dynamic" quantity, in that it is a representative measure of the competing effects of the enthalpic and entropic driving forces involved in a thermodynamic process.

The temperature dependence of the Gibbs energy for an ideal gas is given by the Gibbs–Helmholtz equation, and its pressure dependence is given by,

$$\frac{G}{N} = \frac{G^{\circ}}{N} + kT \ln \frac{p}{p^{\circ}}.$$

If the volume is known rather than pressure, then it becomes,

$$\frac{G}{N} = \frac{G^{\circ}}{N} + kT \ln \frac{V^{\circ}}{V},$$

or more conveniently as its chemical potential:

$$\frac{G}{N} = \mu = \mu^{\circ} + kT \ln \frac{p}{p^{\circ}}.$$

In non-ideal systems, fugacity comes into play.

Derivation

The Gibbs free energy total differential natural variables may be derived by Legendre transforms of the internal energy.

$$\mathrm{d}U = T\mathrm{d}S - p\mathrm{d}V + \sum_i \mu_i \mathrm{d}N_i.$$

The definition of G from above is,

$$G = U + pV - TS.$$

Taking the total differential, we have,

$$\mathrm{d}G = \mathrm{d}U + p\mathrm{d}V + V\mathrm{d}p - T\mathrm{d}S - S\mathrm{d}T.$$

Replacing $\mathrm{d}U$ with the result from the first law gives:

$$\begin{aligned} \mathrm{d}G &= T\mathrm{d}S - p\mathrm{d}V + \sum_i \mu_i \mathrm{d}N_i + p\mathrm{d}V + V\mathrm{d}p - T\mathrm{d}S - S\mathrm{d}T \\ &= V\mathrm{d}p - S\mathrm{d}T + \sum_i \mu_i \mathrm{d}N_i. \end{aligned}$$

The natural variables of G are then p, T, and $\{N_i\}$.

Homogeneous Systems

Because *S*, *V*, and N_i are extensive variables, an Euler integral allows easy integration of d*U*:

$$U = TS - pV + \sum_i \mu_i N_i.$$

Because some of the natural variables of *G* are intensive, d*G* may not be integrated using Euler integrals as is the case with internal energy. However, simply substituting the above integrated result for *U* into the definition of *G* gives a standard expression for *G*:

$$\begin{aligned} G &= U + pV - TS \\ &= (TS - pV + \sum_i \mu_i N_i) + pV - TS \\ &= \sum_i \mu_i N_i. \end{aligned}$$

This result applies to homogeneous, macroscopic systems, but not to all thermodynamic systems.

Gibbs Free Energy of Reactions

The system under consideration is held at constant temperature and pressure, and is closed (no matter can come in or out). The Gibbs energy of any system is $G = U + PV - TS$ and an infinitesimal change in *G*, at constant temperature and pressure yields:

$$dG = dU + PdV - TdS$$

By the first law of thermodynamics, a change in the internal energy *U* is given by:

$$dU = \delta Q + \delta W$$

where δQ is energy added as heat, and δW is energy added as work. The work done on the system may be written as $\delta W = -PdV + \delta W_x$, where $-PdV$ is the mechanical work of compression/expansion done on the system and δW_x is all other forms of work, which may include electrical, magnetic, etc. Assuming that only mechanical work is done,

$$dU = \delta Q - PdV$$

and the infinitesimal change in *G* is:

$$dG = \delta Q - TdS$$

The second law of thermodynamics states that for a closed system, $TdS \geq \delta Q$, and so it follows that:

$$dG \leq 0$$

This means that for a system which is not in equilibrium, its Gibbs energy will always be decreasing, and when it is in equilibrium (i.e. no longer changing), the infinitesimal change dG will be zero. In particular, this will be true if the system is experiencing any number of internal chemical reactions on its path to equilibrium.

In Electrochemical Thermodynamics

When electric charged dQ is passed in an electrochemical cell the emf $\mathcal{E}$ yields a thermodynamic work term that appears in the expression for the change in Gibbs energy:

$$dG = -SdT + VdP + \mathcal{E}dQ,$$

where G is the Gibb's free energy, S is the entropy, V is the system volume, P is its pressure and T is its absolute temperature.

The combination $(\mathcal{E}, Q)$ is an example of a conjugate pair of variables. At constant pressure the above relationship produces a Maxwell relation that links the change in open cell voltage with temperature T (a measurable quantity) to the change in entropy S when charge is passed isothermally and isobarically. The latter is closely related to the reaction entropy of the electrochemical reaction that lends the battery its power. This Maxwell relation is:

$$\left(\frac{\partial \mathcal{E}}{\partial T}\right)_Q = -\left(\frac{\partial S}{\partial Q}\right)_T$$

If a mole of ions goes into solution (for example, in a Daniell cell) the charge through the external circuit is:

$$\Delta Q = -n_0 F_0,$$

where n_0 is the number of electrons/ion, and F_0 is the Faraday constant and the minus sign indicates discharge of the cell. Assuming constant pressure and volume, the thermodynamic properties of the cell are related strictly to the behavior of its emf by:

$$\Delta H = -n_0 F_0 \left(\mathcal{E} - T\frac{d\mathcal{E}}{dT}\right),$$

where ΔH is the enthalpy of reaction. The quantities on the right are all directly measurable.

Free Energy and Temperature

One of the major goals of chemical thermodynamics is to establish criteria for predicting whether a particular reaction or process will occur spontaneously. We have developed one such criterion, the change in entropy of the universe: if $\Delta S_{univ} > 0$ for a process or a reaction, then the process will occur spontaneously as written. Conversely, if $\Delta S_{univ} < 0$, a process cannot occur spontaneously; if $\Delta S_{univ} = 0$, the system is at equilibrium. The sign of ΔS_{univ} is a universally applicable and infallible indicator of the spontaneity of a reaction. Unfortunately, using ΔS_{univ} requires that we calculate ΔS for both a system and its surroundings. This is not particularly useful for two reasons: we are normally much more interested in the system than in the surroundings, and it is difficult to make quantitative measurements of the surroundings (i.e., the rest of the universe). A criterion of spontaneity that is based solely on the state functions of a system would be much more convenient and is provided by a new state function: the Gibbs free energy.

Gibbs Energy and Spontaneity

The criterion for predicting spontaneity is based on (ΔG), the change in G, at constant temperature and pressure. Although very few chemical reactions actually occur under conditions of constant temperature and pressure, most systems can be brought back to the initial temperature and pressure without significantly affecting the value of thermodynamic state functions such as G. At constant temperature and pressure,

$$\Delta G = \Delta H - T\Delta S$$

where all thermodynamic quantities are those of the system. Recall that at constant pressure, $\Delta H = q$, whether a process is reversible or irreversible, and $T\Delta S = q_{rev}$. Using these expressions, we can reduce Equation $\Delta G = \Delta H - T\Delta S$ to $\Delta G = q - q_{rev}$. Thus ΔG is the difference between the heat released during a process (via a reversible or an irreversible path) and the heat released for the same process occurring in a reversible manner. Under the special condition in which a process occurs reversibly, $q = q_{rev}$ and $\Delta G = 0$. As we shall soon see, if ΔG is zero, the system is at equilibrium, and there will be no net change.

What about processes for which $\Delta G \neq 0$? To understand how the sign of ΔG for a system determines the direction in which change is spontaneous, we can rewrite the relationship between ΔS and q_{rev}.

$$\Delta S = \frac{q_{rev}}{T}$$

with the definition of ΔH in terms of q_{rev},

$$q_{rev} = \Delta H$$

to obtain,

$$\Delta S_{surr} = -\frac{\Delta H_{sys}}{t}$$

Thus the entropy change of the surroundings is related to the enthalpy change of the system. We have stated that for a spontaneous reaction, $\Delta S_{univ} > 0$, so substituting we obtain,

$$\Delta S_{univ} = \Delta S_{sys} + \Delta S_{surr} > 0$$
$$= \Delta S_{sys} - \frac{\Delta H_{sys}}{T} > 0$$

Multiplying both sides of the inequality by –T reverses the sign of the inequality; rearranging,

$$\Delta H_{sys} - T\Delta S_{sys} < 0$$

which is equal to ΔG (Equation $\Delta G = \Delta H - T\Delta S$). We can therefore see that for a spontaneous process, $\Delta G < 0$.

The relationship between the entropy change of the surroundings and the heat gained or lost by the system provides the key connection between the thermodynamic properties of the system and the change in entropy of the universe. The relationship shown in Equation $\Delta G = \Delta H - T\Delta S$ allows us to predict spontaneity by focusing exclusively on the thermodynamic properties and temperature of the system. We predict that highly exothermic processes ($\Delta H \ll 0$) that increase the disorder of a system ($\Delta S_{sys} \gg 0$) would therefore occur spontaneously. An example of such a process is the decomposition of ammonium nitrate fertilizer. Ammonium nitrate was also used to destroy the Murrah Federal Building in Oklahoma City, Oklahoma, in 1995. For a system at constant temperature and pressure, we can summarize the following results:

- If $\Delta G < 0$, the process occurs spontaneously.
- If $\Delta G = 0$, the system is at equilibrium.
- If $\Delta G > 0$, the process is not spontaneous as written but occurs spontaneously in the reverse direction.

To further understand how the various components of ΔG dictate whether a process occurs spontaneously, we now look at a simple and familiar physical change: the conversion of liquid water to water vapor. If this process is carried out at 1 atm and the normal boiling point of 100.00°C (373.15 K), we can calculate ΔG from the experimentally measured value of ΔH_{vap} (40.657 kJ/mol). For vaporizing 1 mol of water, $\Delta H = 40{,}657J$, so the process is highly endothermic. From the definition of ΔS (Equation $\Delta S_{surr} = -\frac{\Delta H_{sys}}{T}$), we know that for 1 mol of water,

$$\Delta S_{vap} = \frac{\Delta H_{vap}}{T_b} = \frac{40{,}657}{J373.15}K = 108.96 \text{ J/K}$$

Hence there is an increase in the disorder of the system. At the normal boiling point of water,

$$\Delta G_{100^\circ C} = \Delta H_{100^\circ C} - T\Delta S_{100^\circ C} = 40,657\ \text{J} - [(373.15\ \text{K})(108.96\ \text{J/K})] = 0\ \text{J}$$

The energy required for vaporization offsets the increase in disorder of the system. Thus ΔG = 0, and the liquid and vapor are in equilibrium, as is true of any liquid at its boiling point under standard conditions.

Now suppose we were to superheat 1 mol of liquid water to 110°C. The value of ΔG for the vaporization of 1 mol of water at 110°C, assuming that ΔH and ΔS do not change significantly with temperature, becomes

$$\Delta G_{110^\circ C} = \Delta H - T\Delta S = 40,657\ \text{J} - [(383.15\,\text{K})(108.96\ \text{J/K})] = -1091\ \text{J}$$

At $110^\circ C,\ \Delta G < 0$, and vaporization is predicted to occur spontaneously and irreversibly.

We can also calculate ΔG for the vaporization of 1 mol of water at a temperature below its normal boiling point—for example, 90°C—making the same assumptions:

$$\Delta G_{90^\circ C} = \Delta H - T\Delta S = 40,657\ \text{J} - [(363.15\,\text{K})(108.96\ \text{J/K})] = 1088\ \text{J}$$

At 90°C, $\Delta G > 0$, and water does not spontaneously convert to water vapor. When using all the digits in the calculator display in carrying out our calculations, $\Delta G_{110^\circ C} = 1090\ J = -\Delta G_{90^\circ C}$, as we would predict.

Equilibrium Conditions

$\Delta G = 0$ only if $\Delta H = T\Delta S$.

We can also calculate the temperature at which liquid water is in equilibrium with water vapor. Inserting the values of ΔH and ΔS into the definition of ΔG (Equation $\Delta G = \Delta H - T\Delta S$), setting ΔG = 0, and solving for T,

$$0\ \text{J} = 40,657\ \text{J} - \text{T}\left(108.96\ \text{J/K}\right)\text{T} = 373.15\,\text{K}$$

Thus ΔG = 0 at T = 373.15 K and 1 atm, which indicates that liquid water and water vapor are in equilibrium; this temperature is called the normal boiling point of water. At temperatures greater than 373.15 K, ΔG is negative, and water evaporates spontaneously and irreversibly. Below 373.15 K, ΔG is positive, and water does not evaporate spontaneously. Instead, water vapor at a temperature less than 373.15 K and 1 atm will spontaneously and irreversibly condense to liquid water. Figure shows how the ΔH and TΔS terms vary with temperature for the vaporization of water. When the two lines cross, ΔG = 0, and ΔH = TΔS.

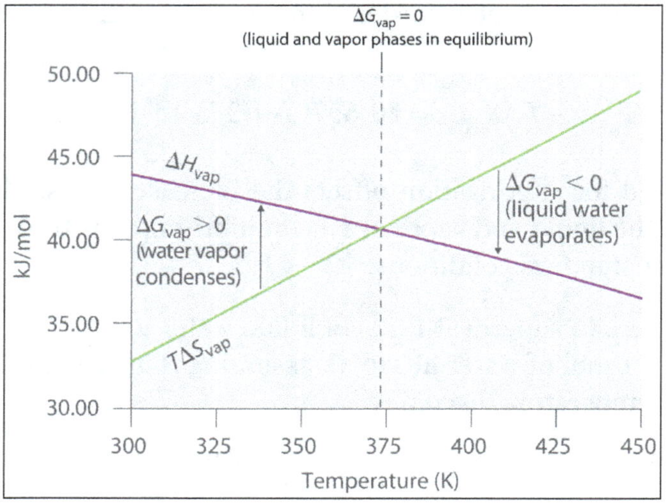

Temperature Dependence of ΔH and TΔS for the Vaporization of Water. Both ΔH and TΔS are temperature dependent, but the lines have opposite slopes and cross at 373.15 K at 1 atm, where ΔH = TΔS. Because ΔG = ΔH − TΔS, at this temperature ΔG = 0, indicating that the liquid and vapor phases are in equilibrium. The normal boiling point of water is therefore 373.15 K. Above the normal boiling point, the TΔS term is greater than ΔH, making ΔG < 0; hence, liquid water evaporates spontaneously. Below the normal boiling point, the ΔH term is greater than TΔS, making ΔG > 0. Thus liquid water does not evaporate spontaneously, but water vapor spontaneously condenses to liquid.

A similar situation arises in the conversion of liquid egg white to a solid when an egg is boiled. The major component of egg white is a protein called albumin, which is held in a compact, ordered structure by a large number of hydrogen bonds. Breaking them requires an input of energy ($\Delta H > 0$), which converts the albumin to a highly disordered structure in which the molecules aggregate as a disorganized solid ($\Delta S > 0$). At temperatures greater than 373 K, the TΔS term dominates, and ΔG < 0, so the conversion of a raw egg to a hard-boiled egg is an irreversible and spontaneous process above 373 K.

ΔH	ΔS	−TΔS	ΔG
-(exothermic)	+(products more disordered)	-(favors spontaneity)	-(spontaneous at all T)
-(exothermic)	-(products less disordered)	+(opposes spontaneity)	-(spontaneous) at low T+(non-spontaneous) at high T "Enthalpically-driven process"
+(endothermic)	+(products more disordered)	-(favors spontaneity)	+(non-spontaneous) at low T-(spontaneous) at high T "Entropically-driven process"
+(endothermic)	-(products less disordered)	+(opposes spontaneity)	+(non-spontaneous at all T)

References

- Chris Vuille; Serway, Raymond A.; Faughn, Jerry S. (2009). College physics. Belmont, CA: Brooks/Cole, Cengage Learning. P. 355. ISBN 978-0-495-38693-3
- Spontaneous-and-nonspontaneous-processes, introchem: lumenlearning.com, Retrieved 26 July, 2020
- Lebon, G., Jou, D., Casas-Vázquez, J. (2008). Understanding Non-equilibrium Thermodynamics. Foundations, Applications, Frontiers, Springer, Berlin, ISBN 978-3-540-74252-4
- Van Ness, Hendrick C. (2003). "H Is for Enthalpy". Journal of Chemical Education. 80 (6): 486. Bibcode:2003jched..80..486V. Doi:10.1021/ed080p486.1
- Zumdahl, Steven S. (2008). "Thermochemistry". Chemistry. Cengage Learning. P. 243. ISBN 978-0-547-12532-9. Archived from the original on 2013-11-14
- Ligrone, Roberto (2019). "Glossary". Biological Innovations that Built the World: A Four-billion-year Journey through Life & Earth History. Entropy. Springer. P. 478. ISBN 3030160564. Retrieved 29 August 2019
- Belkin, Andrey; et., al. (2015). "Self-assembled wiggling nano-structures and the principle of maximum entropy production". Scientific Reports. 5: 8323. Bibcode:2015natsr...5E8323B. Doi:10.1038/srep08323. PMC 4321171. PMID 25662746

Understanding Quantum Chemistry 5

- Electronic Structure
- Electronic Structure Theory
- Atomic Orbital
- Quantum Numbers
- Schrödinger Equation
- Klopman-Salem Equation
- Born–Oppenheimer Approximation

The branch of chemistry which primarily focuses on the applications of quantum mechanics in physical models and chemical systems is referred to as quantum chemistry. Quantum numbers, atomic orbitals, electronic structures, Schrödinger equation, Klopman-Salem equation, etc. are studied within it. This chapter sheds light on the subject of quantum chemistry for a thorough understanding of it.

Quantum chemistry is the application of quantum mechanical principles and equations to the study of molecules. In order to understand matter at its most fundamental level, we must use quantum mechanical models and methods. There are two aspects of quantum mechanics that make it different from previous models of matter.

The first is the concept of wave-particle duality; that is, the notion that we need to think of very small objects (such as electrons) as having characteristics of both particles and waves. Second, quantum mechanical models correctly predict that the energy of atoms and molecules is always quantized, meaning that they may have only specific amounts of energy. Quantum chemical theories allow us to explain the structure of the periodic table, and quantum chemical calculations allow us to accurately predict the structures of molecules and the spectroscopic behavior of atoms and molecules.

Wave Particle Duality for Light

Quantum mechanical ideas began with studies of the physics of light. By the late nineteenth century, virtually all scientists believed that light behaved as a wave. Although some earlier scientists, such as Isaac Newton in the seventeenth century, had thought of light as consisting of particles, the early nineteenth-century experiments of Thomas Young and Augustin Fresnel demonstrated that light has wavelike properties. In these experiments, light was passed through a pair of slits in a screen, and produced alternating light and dark regions (interference patterns) on a second screen. This phenomenon, known as diffraction, cannot be explained using a particle model for light. In the late nineteenth century, James Clerk Maxwell derived a set of equations based on the wave model for light, which beautifully explained most experimental results.

Despite this apparent certainty that light was a wave, Max Planck and Albert Einstein, at the beginning of the twentieth century, showed that some experiments required the use of a particle model for light, rather than a wave model. Since both models were necessary for an accurate description of all of the properties of light, scientists today use mathematical equations appropriate to both waves and particles in describing the properties of light.

Waves and particles are fundamentally different: a particle exists at a particular point in space, whereas a wave continues on for (sometimes) a great distance. It defies intuition to think that both of these models might describe the same thing. Nevertheless, an accurate description of light requires the use of both wave and particle ideas.

The Wave Nature of Matter

The success of wave-particle duality in describing the properties of light paved the way for using that same idea in describing matter. Experiments in the early twentieth century showed that the energy in atoms is quantized; that is, a given atom can have only specific amounts of energy. For hydrogen, the simplest of the atoms, an accurate formula for the possible energies had been experimentally determined but was unexplainable using any particle model for the atom. The best picture that the particle model could give, consistent with experiments on atoms, put the electron in a sort of "orbit" around the nucleus. Unfortunately, the particle model predicts that the electron should collide with the nucleus, releasing energy in the process. Obviously there was a need for a different model for the electron.

One example of quantization would be to think of your distance from the ground when standing on a ladder—your distance from the ground can only change one rung at a time.

In 1924 Louis de Broglie presented a theory for the hydrogen atom that modeled the electron as a wave. Calculations made for this model give the quantization of energy that is experimentally observed in this atom. De Broglie also postulated a general formula for obtaining the wavelength of a moving object. His formula, which is analogous to that used for light, states that the wavelength of a moving object is inversely proportional to its momentum (mass times velocity). When one uses de Broglie's formula to determine wavelengths of macroscopic objects, one discovers that the wavelengths of even the smallest objects visible to the naked eye are too small for the wavelike characteristics of these objects to be significant in any real situation. For the electron, however, the wavelength is large enough to be measurable. Diffraction experiments have been performed using electrons, demonstrating conclusively that they have wave properties.

It is contrary to our intuition that electrons might behave as waves. The repercussions of this notion are that the electron does not have a definite size, but is spread out over a region in space. We are more comfortable with the thought of the electron being a microscopic particle, moving around in an orbit near the nucleus of an atom. As with light, however, we do not abandon the particle model for electrons; rather, we employ mathematical equations arising from both particle and wave models. For quantum chemical calculations, the wave model turns out to be more useful.

The Heisenberg Uncertainty Principle

One consequence of the wave nature of matter is that the position and momentum of small objects are not well known, as they would be for a particle model. In some circumstances, a wave may be confined to a very narrow region in space; however, there is still some uncertainty as to its position. Additionally, the value of the momentum of a quantum object is often not known precisely. In 1927 Werner Heisenberg showed that the product of the uncertainty in position and the uncertainty in momentum is greater than or equal to a certain constant (Planck's constant divided by 4π). This constant is very small; accordingly, quantum mechanical uncertainty in position and momentum of objects that are large enough to see is not noticed experimentally. For electrons, however, quantum mechanical uncertainties in position and momentum are important considerations in interpreting both theoretical models and experimental results. The relationship between the uncertainties in position and momentum is known as the Heisenberg Uncertainty Principle. It tells us that the more we know about the position of a small object, such as an electron, the less we know about its momentum (and vice versa).

Calculating the Wave Function

For a scientist, knowing that matter behaves as a wave is useful only if one knows something about that wave. The wavefunction is a mathematical function describing the wave. For example, $y(x) = \mathrm{A}\sin(kx)$ might be the wave-function for a one-dimensional wave, which exists along the x -axis. Matter waves are three-dimensional; the relevant wavefunction depends on the x, y, and z coordinates of the system being studied (and sometimes on time as well). We conventionally label the wavefunction for a three-dimensional object as $\psi\ (x, y, z)$. In 1926 Erwin Schrödinger introduced a mathematical equation whereby, if one knows the potential energy acting on an object, one can calculate the wavefunction for that object. Heisenberg had already introduced a mathematical formalism for performing quantum mechanics calculations, without explicitly including the concept of waves. It was later shown that, although the approaches of Schrödinger and Heisenberg looked very different, they made exactly the same predictions. In practice, the Schrödinger formalism is more useful for explaining the problem being studied, and the Heisenberg methodology allows for more straightforward computation. Accordingly, a mixture of the two approaches is typically used in modern quantum chemistry. Once we know the wavefunction of the atom or molecule under study, we can calculate the properties of that atom or molecule.

Quantum Mechanics of Atoms

An exact solution for Schrödinger's wave equation can be obtained for the hydrogen atom; however, for larger atoms and molecules (which contain more than one electron), Schrödinger's equation can be solved only approximately. Although this may sound so restrictive as to make the equation useless, there are well-established approaches that allow for practical and accurate calculations on atoms and molecules. This is done by making some assumptions about larger systems based upon the hydrogen atom.

When the Schrödinger equation is solved for the hydrogen atom, the resulting wavefunctions for the various possible energies that the atom can have are used to determine atomic orbitals . An orbital is a region in space where an electron is most likely to be found. For example, the lowest-energy wavefunction for a hydrogen atom is the so-called 1s orbital, which is a spherical region in space surrounding the nucleus. For some higher-energy states, the orbitals are not necessarily spherical in shape (e.g., the 2p orbital pictured).

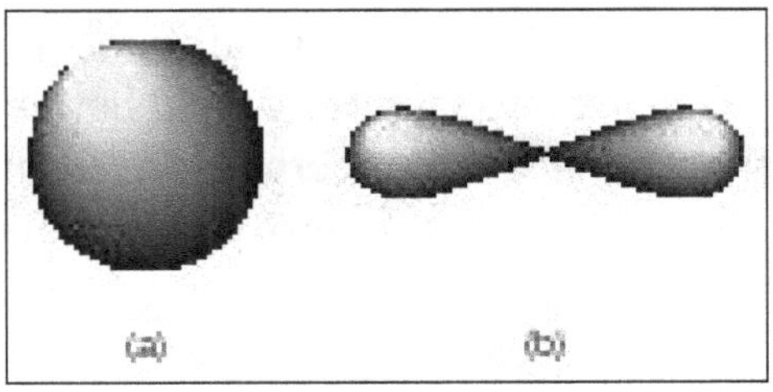

For atoms larger than hydrogen, one assumes that the orbitals occupied by the electrons have the same shape as the hydrogen orbitals but are differing in size and energy. The energies corresponding to these orbitals may be found by solving an approximate version of Schrödinger's equation. These atomic orbitals, in turn, may be used as the building blocks to the electronic behavior in molecules, As it happens, two electrons may share an atomic orbital; we say that these electrons are paired. Chemists have developed a system of rules for determining which orbitals are occupied in which atoms; calculations can then be done to determine the energies of the electrons in the atoms.

For atoms larger than hydrogen, one assumes that the orbitals occupied by the electrons have the same shape as the hydrogen orbitals but are differing in size and energy. The energies corresponding to these orbitals may be found by solving an approximate version of Schrödinger's equation. These atomic orbitals, in turn, may be used as the building blocks to the electronic behavior in molecules, As it happens, two electrons may share an atomic orbital; we say that these electrons are paired. Chemists have developed a system of rules for determining which orbitals are occupied in which atoms; calculations can then be done to determine the energies of the electrons in the atoms.

Quantum Mechanics of Molecules

Molecules are held together by covalent bonds . The simplest definition of a covalent bond is a shared pair of electrons. There are two basic approaches to modeling covalent bonds in molecules: the valence bond model and the molecular orbital model. In the valence bond model, we think of atomic orbitals on each of two atoms combining to form a bond orbital, with one electron from each atom occupying this orbital. Both the bond orbital and the electron pair now "belong" to both of the atoms. This sharing of electrons brings about a lowering in the energy, which makes the formation of molecules from atoms an energetically favorable process. The valence bond model gives the simplest quantum mechanical picture of chemical bonding, but it is not the best method for accurate calculations on molecules containing more than two atoms.

Molecular orbital theory differs from valence bond theory in that it does not require the electrons involved in a bond to be localized between two of the atoms in a molecule. Instead, the electron occupies a molecular orbital, which may be spread out over the entire molecule. As in the valence bond approach, the molecular orbital is formed by adding up contributions from the atomic orbitals on the atoms that make up the molecule. This approach, which does not explicitly model bonds as existing between two atoms, is somewhat less appealing to the intuition than the valence bond approach. However, molecular orbital calculations typically yield better predictions of molecular structure and properties than valence bond methods. Accordingly, most commercially available quantum chemistry software packages rely on molecular orbital methods to perform calculations.

Electronic Structure

In quantum chemistry, electronic structure is the state of motion of electrons in an electrostatic field created by stationary nuclei. The term encompass both the wave functions of the electrons and the energies associated with them. Electronic structure is obtained by solving quantum mechanical equations for the aforementioned clamped-nuclei problem.

Electronic structure problems arise from the Born–Oppenheimer approximation. Along with nuclear dynamics, the electronic structure problem is one of the two steps in studying the quantum mechanical motion of a molecular system. Except for a small number of simple problems such as hydrogen-like atoms, the solution of electronic structure problems require modern computers.

Electronic structure problem is routinely solved with quantum chemistry computer programs. Electronic structure calculations rank among the most computationally intensive tasks in all scientific calculations. For this reason, quantum chemistry calculations take up significant shares on many scientific supercomputer facilities.

A number of methods to obtain electronic structures exist and their applicability varies from case to case.

Electronic Structure Theory

Electronic Structure Theory describes the motions of electrons in atoms or molecules. Generally this is done in the context of the Born-Oppenheimer Approximation, which says that electrons are so much lighter (and therefore faster) than nuclei that they will find their optimal distribution for any given nuclear configuration. The electronic energy at each nuclear configuration is the potential energy that the nuclei feel, so solving the electronic problem for a range of nuclear configurations gives the potential energy surface.

Because the electrons are so small, one needs to use quantum mechanics to solve for their motion. Quantum mechanics tells us that the electrons will not be localized at particular points in space, but they are best thought of as "matter waves" which can interfere. The probability of finding a single electron at a given point in space is given by $\Psi^*(x)\Psi(x)$ for its wavefunction Ψ at the point x. The wavefunction can be determined by solving the time-independent Schrödinger equation $\hat{H}\Psi = E\Psi$. If the problem is time-dependent, then the time-dependent Schrödinger equation $i\hbar\frac{\partial\Psi}{\partial t} = \hat{H}\Psi$ must be used instead; otherwise, the solutions to the time-independent problem are also solutions to the time-dependent problem when they are multiplied by the energy dependent phase factor $e^{-iEt/\hbar}$. Since we have fixed the nuclei under the Born-Oppenheimer approximation, we solve for the nonrelativistic *electronic* Schrödinger equation:

$$\hat{H} = -\frac{\hbar^2}{2m}\sum_i \nabla_i^2 - \sum_A \frac{\hbar^2}{2M_2}\nabla_A^2 - \sum_{A,i}\frac{Z_A e^2}{4\pi \in_0 r_{Ai}} + \sum_{A>B}\frac{Z_A Z_B e^2}{4\pi \in_0 R_{AB}} + \sum_{i>j}\frac{e^2}{4\pi \in_0 r_{ij}}$$

where i, j refer to electrons and A, B refer to nuclei. In atomic units, this simplifies to;

$$\hat{} = \frac{1}{2}\sum_{i}\nabla_i^2 - \sum_{A}\frac{}{2M}\nabla_A^2 \quad \sum_{A\,i}\frac{Z_A}{r_{Ai}} + \sum_{A>B}\frac{Z_A Z_B}{R_{AB}} + \sum_{i>j}\frac{1}{r_{ij}}$$

This Hamiltonian is appropriate as long as relativistic effects are not important for the system in question. Roughly speaking, relativistic effects are not generally considered important for atoms with atomic number below about 25 (Mn). For heavier atoms, the inner electrons are held more tightly to the nucleus and have velocities which increase as the atomic number increases; as these velocities approach the speed of light, relativistic effects become more important. There are various approaches for accounting for relativistic effects, but the most popular is to use relativistic effective core potentials (RECPs), often along with the standard nonrelativistic Hamiltonian.

Valence Bond Theory

In chemistry, valence bond (VB) theory is one of the two basic theories, along with molecular orbital (MO) theory, that were developed to use the methods of quantum mechanics to explain chemical bonding. It focuses on how the atomic orbitals of the dissociated atoms combine to give individual chemical bonds when a molecule is formed. In contrast, molecular orbital theory has orbitals that cover the whole molecule.

According to this theory a covalent bond is formed between two atoms by the overlap of half filled valence atomic orbitals of each atom containing one unpaired electron. A valence bond structure is similar to a Lewis structure, but where a single Lewis structure cannot be written, several valence bond structures are used. Each of these VB structures represents a specific Lewis structure. This combination of valence bond structures is the main point of resonance theory. Valence bond theory considers that the overlapping atomic orbitals of the participating atoms form a chemical bond. Because of the overlapping, it is most probable that electrons should be in the bond region. Valence bond theory views bonds as weakly coupled orbitals (small overlap). Valence bond theory is typically easier to employ in ground state molecules. The inner-shell orbitals and electrons remain essentially unchanged during the formation of bonds.

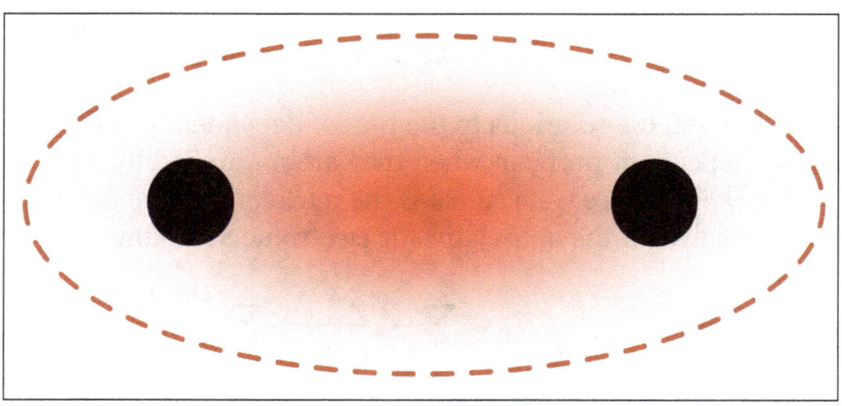

σ bond between two atoms: localization of electron density.

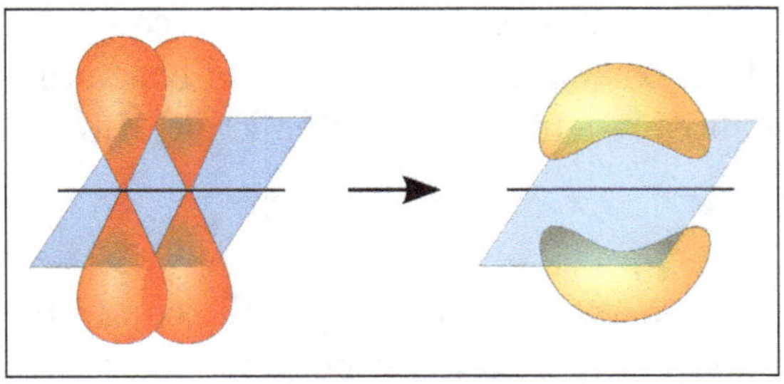

Two p-orbitals forming a π-bond.

The overlapping atomic orbitals can differ. The two types of overlapping orbitals are sigma and pi. Sigma bonds occur when the orbitals of two shared electrons overlap head-to-head. Pi bonds occur when two orbitals overlap when they are parallel. For example, a bond between two *s*-orbital electrons is a sigma bond, because two spheres are always coaxial. In terms of bond order, single bonds have one sigma bond, double bonds consist of one sigma bond and one pi bond, and triple bonds contain one sigma bond and two pi bonds. However, the atomic orbitals for bonding may be hybrids. Often, the bonding atomic orbitals have a character of several possible types of orbitals. The methods to get an atomic orbital with the proper character for the bonding is called hybridization.

Modern valence bond theory now complements molecular orbital theory, which does not adhere to the valence bond idea that electron pairs are localized between two specific atoms in a molecule but that they are distributed in sets of molecular orbitals which can extend over the entire molecule. Molecular orbital theory can predict magnetic and ionization properties in a straightforward manner, while valence bond theory gives similar results but is more complicated. Modern valence bond theory views aromatic properties of molecules as due to spin coupling of the π orbitals. This is essentially still the old idea of resonance between Kekulé and Dewar structures. In contrast, molecular orbital theory views aromaticity as delocalization of the π-electrons. Valence bond treatments are restricted to relatively small molecules, largely due to the lack of orthogonality between valence bond orbitals and between valence bond structures, while molecular orbitals are orthogonal. On the other hand, valence bond theory provides a much more accurate picture of the reorganization of electronic charge that takes place when bonds are broken and formed during the course of a chemical reaction. In particular, valence bond theory correctly predicts the dissociation of homonuclear diatomic molecules into separate atoms, while simple molecular orbital theory predicts dissociation into a mixture of atoms and ions. For example, the molecular orbital function for dihydrogen is an equal mixture of the covalent and ionic valence bond structures and so predicts incorrectly that the molecule would dissociate into an equal mixture of hydrogen atoms and hydrogen positive and negative ions.

Modern valence bond theory replaces the overlapping atomic orbitals by overlapping valence bond orbitals that are expanded over a large number of basis functions, either centered each on one atom to give a classical valence bond picture, or centered on all atoms in the molecule. The resulting energies are more competitive with energies from calculations where electron correlation is introduced based on a Hartree–Fock reference wavefunction. The most recent text is by Shaik and Hiberty.

Applications

An important aspect of the valence bond theory is the condition of maximum overlap, which leads to the formation of the strongest possible bonds. This theory is used to explain the covalent bond formation in many molecules.

For example, in the case of the F_2 molecule, the F–F bond is formed by the overlap of p_z orbitals of the two F atoms, each containing an unpaired electron. Since the nature of the overlapping orbitals are different in H_2 and F_2 molecules, the bond strength and bond lengths differ between H_2 and F_2 molecules.

In an HF molecule the covalent bond is formed by the overlap of the 1*s* orbital of H and the $2p_z$ orbital of F, each containing an unpaired electron. Mutual sharing of electrons between H and F results in a covalent bond in HF.

Molecular Orbital Theory

In chemistry, Molecular orbital (MO) theory is a method for describing the electronic structure of molecules using quantum mechanics. Electrons are not assigned to individual bonds between atoms, but are treated as moving under the influence of the nuclei in the whole molecule. The spatial and energetic properties of electrons are described by quantum mechanics as molecular orbitals surround two or more atoms in a molecule and contain valence electrons between atoms. Molecular orbital theory, which was proposed in the early twentieth century, revolutionized the study of bonding by approximating the states of bonded electrons—the molecular orbitals—as linear combinations of atomic orbitals (LCAO). These approximations are now made by applying the density functional theory (DFT) or Hartree–Fock (HF) models to the Schrödinger equation.

The other foundational theory of quantum chemistry is the valence bond theory.

MO theory provides a global, delocalized perspective on chemical bonding. In MO theory, *any* electron in a molecule may be found *anywhere* in the molecule, since quantum conditions allow electrons to travel under the influence of an arbitrarily large number of nuclei, as long as they are in eigenstates permitted by certain quantum rules. Thus, when excited with the requisite amount of energy through high-frequency light or other means, electrons can transition to higher-energy molecular orbitals. For instance, in the simple case of a hydrogen diatomic molecule, promotion of a single electron from a bonding orbital to an antibonding orbital can occur under UV radiation. This promotion weakens the bond between the two hydrogen atoms and can lead to photodissociation—the breaking of a chemical bond due to the absorption of light.

Although in MO theory *some* molecular orbitals may hold electrons that are more localized between specific pairs of molecular atoms, *other* orbitals may hold electrons that are spread more uniformly over the molecule. Thus, overall, bonding is far more delocalized in MO theory, which makes it more applicable to resonant molecules that have equivalent non-integer bond orders than valence bond (VB) theory. This makes MO theory more useful for the description of extended systems.

An example is the MO description of benzene, C_6H_6, which is an aromatic hexagonal ring of six carbon atoms and three double bonds. In this molecule, 24 of the 30 total valence bonding electrons—24 coming from carbon atoms and 6 coming from hydrogen atoms—are located in 12 σ (sigma) bonding orbitals, which are located mostly between pairs of atoms (C-C or C-H), similarly to the electrons in the valence bond description. However, in benzene the remaining six bonding electrons are located in three π (pi) molecular bonding orbitals that are delocalized around the ring. Two of these electrons are in an MO that has equal orbital contributions from all six atoms. The other four electrons are in orbitals with vertical nodes at right angles to each other. As in the VB theory, all of these six delocalized π electrons reside in a larger space that exists above and below the ring plane. All carbon-carbon bonds in benzene are chemically equivalent. In MO theory this is a direct consequence of the fact that the three molecular π orbitals combine and evenly spread the extra six electrons over six carbon atoms.

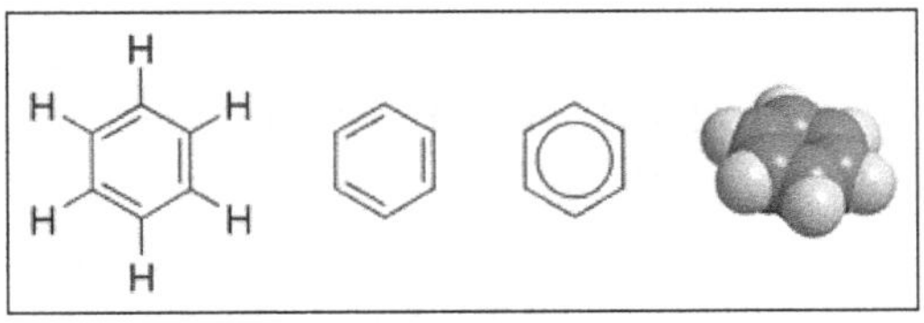

Structure of benzene.

In molecules such as methane, CH_4, the eight valence electrons are found in four MOs that are spread out over all five atoms. However, it is possible to transform the MOs into four localized sp^3 orbitals. Linus Pauling, in 1931, hybridized the carbon 2s and 2p orbitals so that they pointed directly at the hydrogen 1s basis functions and featured maximal overlap. However, the delocalized MO description is more appropriate for predicting ionization energies and the positions of spectral absorption bands. When methane is ionized, a single electron is taken from the valence MOs, which can come from the s bonding or the triply degenerate p bonding levels, yielding two ionization energies. In comparison, the explanation in VB theory is more complicated. When one electron is removed from an sp^3 orbital, resonance is invoked between four valence bond structures, each of which has a single one-electron bond and three two-electron bonds. Triply degenerate T_2 and A_1 ionized states (CH_4^+) are produced from different linear combinations of these four structures. The difference in energy between the ionized and ground state gives the two ionization energies.

As in benzene, in substances such as beta carotene, chlorophyll, or heme, some electrons in the π orbitals are spread out in molecular orbitals over long distances in a molecule, resulting in light absorption in lower energies (the visible spectrum), which accounts for the characteristic colours of these substances. This and other spectroscopic data for molecules are well explained in MO theory, with an emphasis on electronic states associated with multicenter orbitals, including mixing of orbitals premised on principles of orbital symmetry matching. The same MO principles also naturally explain some electrical phenomena, such as high electrical conductivity in the planar direction of the hexagonal atomic sheets that exist in graphite. This results from continuous band overlap of half-filled p orbitals and explains electrical conduction. MO theory recognizes that some electrons in the graphite atomic sheets are completely delocalized over arbitrary distances, and reside in very large molecular orbitals that cover an entire graphite sheet, and some electrons are thus as free to move and therefore conduct electricity in the sheet plane, as if they resided in a metal.

Linear Combination of Atomic Orbitals (LCAO) Method

In the LCAO method, each molecule has a set of molecular orbitals. It is assumed that the molecular orbital wave function ψ_j can be written as a simple weighted sum of the n constituent atomic orbitals χ_i, according to the following equation:

$$\psi_j = \sum_{i=1}^{n} c_{ij}\chi_i.$$

One may determine c_{ij} coefficients numerically by substituting this equation into the Schrödinger equation and applying the variational principle. The variational principle is a mathematical technique used in quantum mechanics to build up the coefficients of each atomic orbital basis. A larger coefficient means that the orbital basis is composed more of that particular contributing atomic orbital—hence, the molecular orbital is best characterized by that type. This method of quantifying orbital contribution as a linear combination of atomic orbitals is used in computational chemistry. An additional unitary transformation can be applied on the system to accelerate the convergence in some computational schemes. Molecular orbital theory was seen as a competitor to valence bond theory in the 1930s, before it was realized that the two methods are closely related and that when extended they become equivalent.

Types of Orbitals

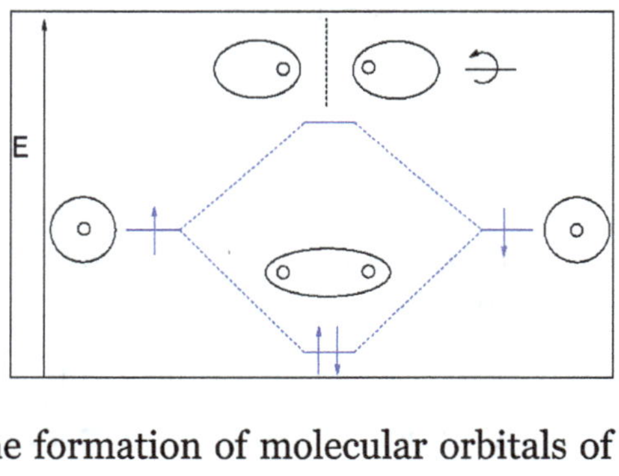

MO diagram showing the formation of molecular orbitals of H_2 (centre) from atomic orbitals of two H atoms. The lower-energy MO is bonding with electron density concentrated between the two H nuclei. The higher-energy MO is anti-bonding with electron density concentrated behind each H nucleus.

Molecular orbital (MO) theory uses a linear combination of atomic orbitals (LCAO) to represent molecular orbitals resulting from bonds between atoms. These are often divided into three types, bonding, antibonding, and non-bonding. A bonding orbital concentrates electron density in the region *between* a given pair of atoms, so that its electron density will tend to attract each of the two nuclei toward the other and hold the two atoms together. An anti-bonding orbital concentrates electron density "behind" each nucleus (i.e. on the side of each atom which is farthest from the other atom), and so tends to pull each of the two nuclei away from the other and actually weaken the bond between the two nuclei. Electrons in non-bonding orbitals tend to be associated with atomic orbitals that do not interact positively or negatively with one another, and electrons in these orbitals neither contribute to nor detract from bond strength.

Molecular orbitals are further divided according to the types of atomic orbitals they are formed from. Chemical substances will form bonding interactions if their orbitals become lower in energy when they interact with each other. Different bonding orbitals are distinguished that differ by electron configuration (electron cloud shape) and by energy levels.

The molecular orbitals of a molecule can be illustrated in molecular orbital diagrams.

Atomic Orbital

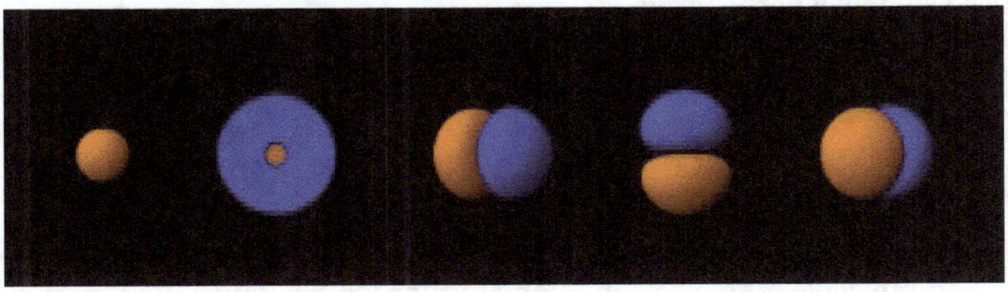

The shapes of the first five atomic orbitals are: 1s, 2s, $2p_x$, $2p_y$, and $2p_z$. The two colors show the phase or sign of the wave function in each region. Each picture is domain coloring of a $\psi(x, y, z)$ function which depend on the coordinates of one electron.

In atomic theory and quantum mechanics, an atomic orbital is a mathematical function that describes the wave-like behavior of either one electron or a pair of electrons in an atom. This function can be used to calculate the probability of finding any electron of an atom in any specific region around the atom's nucleus. The term *atomic orbital* may also refer to the physical region or space where the electron can be calculated to be present, as defined by the particular mathematical form of the orbital.

Each orbital in an atom is characterized by a unique set of values of the three quantum numbers n, ℓ, and m, which respectively correspond to the electron's energy, angular momentum, and an angular momentum vector component (the magnetic quantum number). Each such orbital can be occupied by a maximum of two electrons, each with its own spin quantum number s. The simple names s orbital, p orbital, d orbital and f orbital refer to orbitals with angular momentum quantum number $\ell = 0, 1, 2$ and 3 respectively. These names, together with the value of n, are used to describe the electron configurations of atoms. They are derived from the description by early spectroscopists of certain series of alkali metal spectroscopic lines as sharp, principal, diffuse, and fundamental. Orbitals for $\ell > 3$ continue alphabetically, omitting j (g, h, i, k, ...) because some languages do not distinguish between the letters "i" and "j".

Atomic orbitals are the basic building blocks of the atomic orbital model (alternatively known as the electron cloud or wave mechanics model), a modern framework for visualizing the submicroscopic behavior of electrons in matter. In this model the electron cloud of a multi-electron atom may be seen as being built up (in approximation) in an electron configuration that is a product of simpler hydrogen-like atomic orbitals. The repeating *periodicity* of the blocks of 2, 6, 10, and 14 elements within sections of the periodic table arises naturally from the total number of electrons that occupy a complete set of s, p, d and f atomic orbitals, respectively, although for higher values of the quantum number n, particularly when the atom in question bears a positive charge, the energies of certain sub-shells become very similar and so the order in which they are said to be populated by electrons (e.g. Cr = $[Ar]4s^1 3d^5$ and Cr^{2+} = $[Ar]3d^4$) can only be rationalized somewhat arbitrarily.

Quantum Numbers

A quantum number is a value that is used when describing the energy levels available to atoms and molecules. An electron in an atom or ion has four quantum numbers to describe its state and yield solutions to the Schrödinger wave equation for the hydrogen atom.

There are four quantum numbers:

- n - principal quantum number - describes the energy level.
- ℓ - azimuthal or angular momentum quantum number - describes the subshell.

- m_ℓ or m - magnetic quantum number - describes the orbital of the subshell.
- m_s or s - spin quantum number - describes the spin.

Principal Quantum Number

In quantum mechanics, the principal quantum number (symbolized n) is one of four quantum numbers assigned to all electrons in an atom to describe that electron's state. Its values are natural numbers (from 1) making it a discrete variable.

Apart from the principal quantum number, the other quantum numbers for bound electrons are the azimuthal quantum number ℓ, the magnetic quantum number m_l, and the spin quantum numbers.

As n increases, the electron is also at a higher energy and is, therefore, less tightly bound to the nucleus. For higher n the electron is, in average, farther from the nucleus. For each value of n there are n accepted ℓ (azimuthal) values ranging from 0 to $n - 1$ inclusively, hence higher-n electron states are more numerous. Accounting for two states of spin, each n-shell can accommodate up to $2n^2$ electrons.

In a simplistic one-electron model described below, the total energy of an electron is a negative inverse quadratic function of the principal quantum number n, leading to degenerate energy levels for each $n > 1$. In more complex systems—those having forces other than the nucleus–electron Coulomb force—these levels split. For multielectron atoms this splitting results in "subshells" parametrized by ℓ. Description of energy levels based on n alone gradually becomes inadequate for atomic numbers starting from 5 (boron) and fails completely on potassium ($Z = 19$) and afterwards.

The principal quantum number was first created for use in the semiclassical Bohr model of the atom, distinguishing between different energy levels. With the development of modern quantum mechanics, the simple Bohr model was replaced with a more complex theory of atomic orbitals. However, the modern theory still requires the principal quantum number.

Derivation

There is a set of quantum numbers associated with the energy states of the atom. The four quantum numbers n, ℓ, m, and s specify the complete and unique quantum state of a single electron in an atom, called its wave function or orbital. Two electrons belonging to the same atom cannot have the same values for all four quantum numbers, due to the Pauli exclusion principle. The wavefunction of the Schrödinger wave equation reduces to the three equations that when solved lead to the first three quantum numbers. Therefore, the equations for the first three quantum numbers are all interrelated. The principal quantum number arose in the solution of the radial part of the wave equation as shown below.

The Schrödinger wave equation describes energy eigenstates with corresponding real numbers E_n and a definite total energy, the value of E_n. The bound state energies of the electron in the hydrogen atom are given by:

$$E_n = \frac{E_1}{n^2} = \frac{-13.6 \text{ eV}}{n^2}, \quad n = 1, 2, 3, \ldots$$

The parameter n can take only positive integer values. The concept of energy levels and notation were taken from the earlier Bohr model of the atom. Schrödinger's equation developed the idea from a flat two-dimensional Bohr atom to the three-dimensional wavefunction model.

In the Bohr model, the allowed orbits were derived from quantized (discrete) values of orbital angular momentum, L according to the equation,

$$\text{L} = n \cdot \hbar = n \cdot \frac{h}{2\pi}$$

where n = 1, 2, 3, ... and is called the principal quantum number, and h is Planck's constant. This formula is not correct in quantum mechanics as the angular momentum magnitude is described by the azimuthal quantum number, but the energy levels are accurate and classically they correspond to the sum of potential and kinetic energy of the electron.

The principal quantum number n represents the relative overall energy of each orbital. The energy level of each orbital increases as its distance from the nucleus increases. The sets of orbitals with the same n value are often referred to as an electron shell.

The minimum energy exchanged during any wave–matter interaction is the product of the wave frequency multiplied by Planck's constant. This causes the wave to display particle-like packets of energy called quanta. The difference between energy levels that have different n determine the emission spectrum of the element.

In the notation of the periodic table, the main shells of electrons are labeled:

$$K(n=1), L(n=2), M(n=3), \text{etc.}$$

based on the principal quantum number.

The principal quantum number is related to the radial quantum number, n_r, by:

$$n = n_r + \ell + 1$$

where ℓ is the azimuthal quantum number and n_r is equal to the number of nodes in the radial wavefunction.

The definite total energy for a particle motion in a common Coulomb field and with a discrete spectrum, is given by:

$$E_n = -\frac{Z^2\hbar^2}{2m_0 a_B^2 n^2} = -\frac{Z^2 e^4 m_0}{2\hbar^2 n^2},$$

where a_B is the Bohr radius.

This discrete energy spectrum resulted from the solution of the quantum mechanical problem on the electron motion in the Coulomb field, coincides with the spectrum that was obtained with the help application of the Bohr–Sommerfeld quantization rules to the classical equations. The radial quantum number determines the number of nodes of the radial wave function $R(r)$.

Values

In chemistry, values $n = 1, 2, 3, 4, 5, 6, 7$ are used in relation to the electron shell theory, with expected inclusion of $n = 8$ (and possibly 9) for yet not accessible period-8 elements. In atomic physics higher n occur for description of excited states. Observations of interstellar medium reveal atomic hydrogen spectral lines involving n on order of hundreds; values up to 766 were detected.

Azimuthal Quantum Number

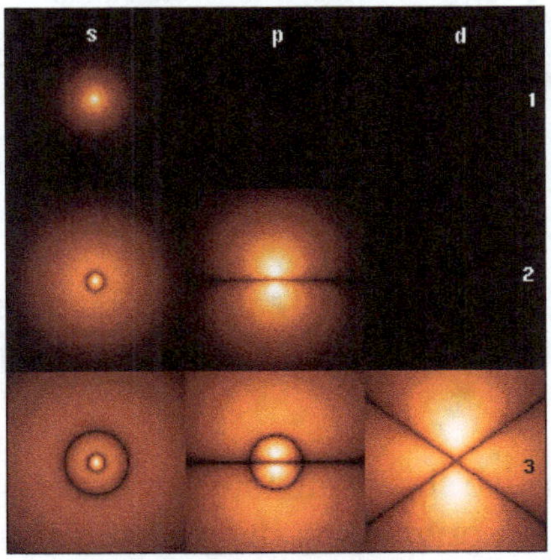

The atomic orbital wavefunctions of a hydrogen atom. The principal quantum number (n) is at the right of each row and the azimuthal quantum number (ℓ) is denoted by letter at top of each column.

The azimuthal quantum number is a quantum number for an atomic orbital that determines its orbital angular momentum and describes the shape of the orbital. The azimuthal quantum number is the second of a set of quantum numbers which describe the unique quantum state of an electron (the others being the principal quantum number, the magnetic quantum number, and the spin quantum number). It is also known as the orbital angular momentum quantum number, orbital quantum number or second quantum number, and is symbolized as ℓ.

Derivation

Connected with the energy states of the atom's electrons are four quantum numbers: n, ℓ, m_ℓ, and m_s. These specify the complete, unique quantum state of a single electron in an atom, and make up its wavefunction or *orbital*. The wavefunction of the Schrödinger equation reduces to three equations that, when solved, lead to the first three quantum numbers. Therefore, the equations for the first three quantum numbers are all interrelated. The azimuthal quantum number arose in the solution of the polar part of the wave equation as shown below, reliant on the spherical coordinate system, which generally works best with models having some glimpse of spherical symmetry.

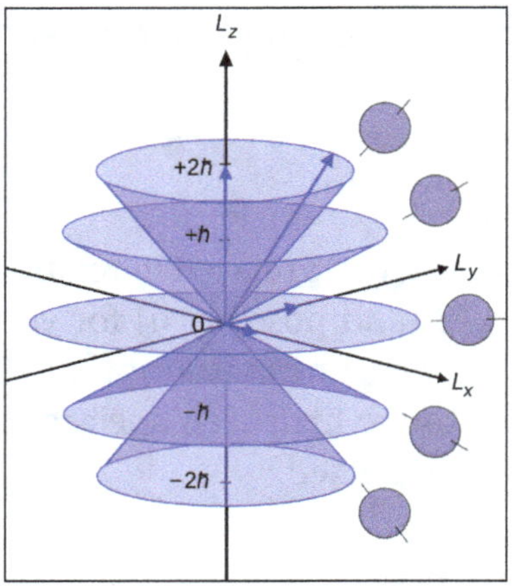

Illustration of quantum mechanical orbital angular momentum.

An atomic electron's angular momentum, L, is related to its quantum number ℓ by the following equation:

$$L^2\,\Psi = \hbar^2 \ell(\ell+1)\Psi$$

where $\hbar$ is the reduced Planck's constant, L^2 is the orbital angular momentum operator and ψ is the wavefunction of the electron. The quantum number ℓ is always a non-negative integer: 0, 1, 2, 3, etc.

Atomic orbitals have distinctive shapes denoted by letters. In the illustration, the letters s, p, and d (a convention originating in spectroscopy) describe the shape of the atomic orbital.

Their wavefunctions take the form of spherical harmonics, and so are described by Legendre polynomials. The various orbitals relating to different values of ℓ are sometimes called sub-shells, and are referred, as follows:

Quantum Subshells for the Azimuthal Quantum Number				
Azimuthal number (ℓ)	Historical Letter	Maximum Electrons	Historical Name	Shape
0	s	2	sharp	spherical

1	p	6	principal	three dumbbell-shaped polar-aligned orbitals; one lobe on each pole of the x, y, and z (+ and – axes)
2	d	10	diffuse	nine dumbbells and one doughnut
3	f	14	funda-mental	"unique shape #2"
4	g	18		
5	h	22		
6	i	26		
The letters after the f sub-shell just follow letter *f* in alphabetical order except the letter *j* and those already used.				

Each of the different angular momentum states can take $2(2\ell + 1)$ electrons. This is because the third quantum number m_ℓ (which can be thought of loosely as the quantized projection of the angular momentum vector on the z-axis) runs from $-\ell$ to ℓ in integer units, and so there are $2\ell + 1$ possible states. Each distinct n, ℓ, m_ℓ orbital can be occupied by two electrons with opposing spins (given by the quantum number $m_s = \pm 1/2$), giving $2(2\ell + 1)$ electrons overall. Orbitals with higher ℓ than given in the table are perfectly permissible, but these values cover all atoms so far discovered.

For a given value of the principal quantum number n, the possible values of ℓ range from 0 to $n - 1$; therefore, the $n = 1$ shell only possesses an s subshell and can only take 2 electrons, the $n = 2$ shell possesses an s and a p subshell and can take 8 electrons overall, the $n = 3$ shell possesses s, p, and d subshells and has a maximum of 18 electrons, and so on.

A simplistic one-electron model results in energy levels depending on the principal number alone. In more complex atoms these energy levels split for all $n > 1$, placing states of higher ℓ above states of lower ℓ. For example, the energy of 2p is higher than of 2s, 3d occurs higher than 3p, which in turn is above 3s, etc. This effect eventually forms the block structure of the periodic table. No known atom possesses an electron having ℓ higher than three (f) in its ground state.

The angular momentum quantum number, ℓ, governs the number of planar nodes going through the nucleus. A planar node can be described in an electromagnetic wave as the midpoint between crest and trough, which has zero magnitude. In an s orbital, no nodes go through the nucleus, therefore the corresponding azimuthal quantum number ℓ takes the value of 0. In a p orbital, one node traverses the nucleus and therefore ℓ has the value of 1. L has the value $\sqrt{2}\hbar$.

Depending on the value of n, there is an angular momentum quantum number ℓ and the following series. The wavelengths listed are for a hydrogen atom:

- $n = 1, L = 0$, Lyman series (ultraviolet).
- $n = 2, L = \sqrt{2}\hbar$, Balmer series (visible).

- $n = 3, L = \sqrt{6}\hbar$, Ritz–Paschen series (near infrared).
- $n = 4, L = 2\sqrt{3}\hbar$, Brackett series (short-wavelength infrared).
- $n = 5, L = 2\sqrt{5}\hbar$, Pfund series (mid-wavelength infrared).

Addition of Quantized Angular Momenta

Given a quantized total angular momentum $\vec{j}$ which is the sum of two individual quantized angular momenta $\overrightarrow{\ell_1}$ and $\overrightarrow{\ell_2}$,

$$\vec{j} = \overrightarrow{\ell_1} + \overrightarrow{\ell_2}$$

the quantum number j associated with its magnitude can range from $|\ell_1 - \ell_2|$ to $\ell_1 + \ell_2$ in integer steps where ℓ_1 and ℓ_2 are quantum numbers corresponding to the magnitudes of the individual angular momenta.

Total Angular Momentum of an Electron in the Atom

"Vector cones" of total angular momentum J (purple), orbital L (blue), and spin S (green). The cones arise due to quantum uncertainty between measuring angular momentum components.

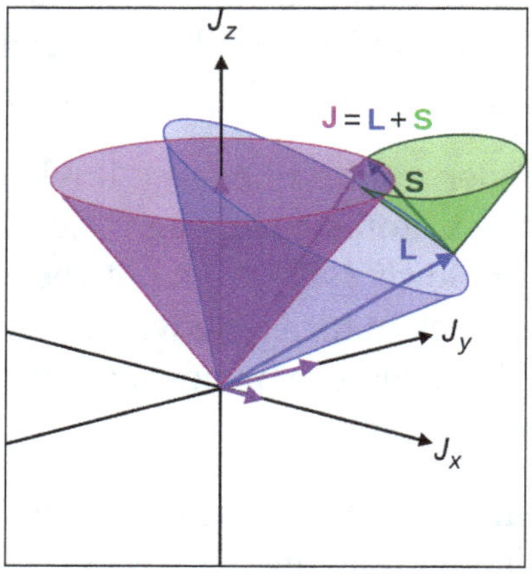

Due to the spin–orbit interaction in the atom, the orbital angular momentum no longer commutes with the Hamiltonian, nor does the spin. These therefore change over time. However the total angular momentum J does commute with the one-electron Hamiltonian and so is constant. J is defined through,

$$\vec{J} = \vec{L} + \vec{S}$$

L being the orbital angular momentum and S the spin. The total angular momentum satisfies the same commutation relations as orbital angular momentum, namely,

$$[J_i, J_j] = i\hbar\epsilon_{ijk}J_k$$

from which follows,

$$\left[J_i, J^2\right] = 0$$

where J_i stand for J_x, J_y, and J_z.

The quantum numbers describing the system, which are constant over time, are now j and mj, defined through the action of J on the wavefunction Ψ

$$J^2\Psi = \hbar^2 j(j+1)\Psi$$

$$J_z\Psi = \hbar m_j \Psi$$

So that *j* is related to the norm of the total angular momentum and m_j to its projection along a specified axis. The *j* number has a particular importance for relativistic quantum chemistry, often featuring in subscript in electron configuration of superheavy elements.

As with any angular momentum in quantum mechanics, the projection of J along other axes cannot be co-defined with J_z, because they do not commute.

Relation between New and Old Quantum Numbers

j and m_j, together with the parity of the quantum state, replace the three quantum numbers ℓ, m_ℓ and m_s (the projection of the spin along the specified axis). The former quantum numbers can be related to the latter.

Furthermore, the eigenvectors of *j*, *s*, m_j and parity, which are also eigenvectors of the Hamiltonian, are linear combinations of the eigenvectors of ℓ, *s*, m_ℓ and m_s.

List of Angular Momentum Quantum Numbers

- Intrinsic (or spin) angular momentum quantum number, or simply spin quantum number.
- Orbital angular momentum quantum number.
- Magnetic quantum number, related to the orbital momentum quantum number.
- Total angular momentum quantum number.

Spin Quantum Number

In atomic physics, the spin quantum number is a quantum number that parameterizes the intrinsic angular momentum (or spin angular momentum, or simply spin) of a given particle. The spin quantum number is the fourth of a set of quantum numbers (the principal quantum number, the azimuthal quantum number, the magnetic quantum number, and the spin quantum number), which completely describe the quantum state of an electron. It is designated by the letter *s*. It describes the energy, shape and orientation of orbitals. The name comes from a physical spinning (denoted by the letter s) about an axis that was proposed by Uhlenbeck and Goudsmit. However this simplistic picture was quickly realized to be physically impossible and replaced by a more abstract quantum-mechanical description.

Derivation

As a solution for a certain partial differential equation, the quantized angular momentum can be written as:

$$\| s \| = \sqrt{s(s+1)}\hbar$$

where:

- s is the quantized spin vector.
- $\| s \|$ is the norm of the spin vector.
- s is the spin quantum number associated with the spin angular momentum.
- $\hbar$ is the reduced Planck constant.

Given an arbitrary direction *z* (usually determined by an external magnetic field) the spin *z*-projection is given by,

$$s_z = m_s \hbar$$

where m_s is the secondary spin quantum number, ranging from $-s$ to $+s$ in steps of one. This generates $2s + 1$ different values of m_s.

The allowed values for *s* are non-negative integers or half-integers. Fermions (such as the electron, proton or neutron) have half-integer values, whereas bosons (e.g., photon, mesons) have integer spin values.

Algebra

The algebraic theory of spin is a carbon copy of the angular momentum in quantum mechanics theory. First of all, spin satisfies the fundamental commutation relation:

$$[S_i, S_j] = i\hbar\epsilon_{ijk}S_k,$$

$$[S_i, S^2] = 0$$

where ϵ_{ijk} is the (antisymmetric) Levi-Civita symbol. This means that it is impossible to know two coordinates of the spin at the same time because of the restriction of the uncertainty principle.

Next, the eigenvectors of S^2 and S_z satisfy:

$$S^2 \,|\, s, m_s\rangle = \hbar^2 s(s+1) \,|\, s, m_s\rangle$$

$$S_z \,|\, s, m_s\rangle = \hbar m_s \,|\, s, m_s\rangle$$

$$S_\pm \,|\, s, m_s\rangle = \hbar\sqrt{s(s+1) - m_s(m_s \pm 1)} \,|\, s, m_s \pm 1\rangle$$

where $S_\pm = S_x \pm iS_y$ are the creation and annihilation (or "raising" and "lowering" or "up" and "down") operators.

Electron Spin

The spin angular momentum is characterized by a quantum number; s = 1/2 specifically for electrons. In a way analogous to other quantized angular momenta, L, it is possible to obtain an expression for the total spin angular momentum:

$$S = \hbar\sqrt{\frac{1}{2}\left(\frac{1}{2}+1\right)} = \frac{\sqrt{3}}{2}\hbar$$

where:

$\hbar$ is the reduced Planck constant.

The hydrogen spectra fine structure is observed as a doublet corresponding to two possibilities for the *z*-component of the angular momentum, where for any given direction *z*:

$$S_z = \pm\frac{1}{2}\hbar$$

whose solution has only two possible *z*-components for the electron. In the electron, the two different spin orientations are sometimes called "spin-up" or "spin-down".

The spin property of an electron would give rise to magnetic moment, which was a requisite for the fourth quantum number. The electron spin magnetic moment is given by the formula:

$$\mu_s = -\frac{e}{2m} gS$$

where:

- *e* is the charge of the electron.
- *g* is the Landé g-factor.

and by the equation:

$$\mu_z = \pm\frac{1}{2} g\mu_B$$

where μ_B is the Bohr magneton.

When atoms have even numbers of electrons the spin of each electron in each orbital has opposing orientation to that of its immediate neighbor(s). However, many atoms have an odd number of electrons or an arrangement of electrons in which there is an unequal number of "spin-up" and "spin-down" orientations. These atoms or electrons are said to have unpaired spins that are detected in electron spin resonance.

Detection of Spin

When lines of the hydrogen spectrum are examined at very high resolution, they are found to be closely spaced doublets. This splitting is called fine structure, and was one of the first experimental evidences for electron spin. The direct observation of the electron's intrinsic angular momentum was achieved in the Stern–Gerlach experiment.

The Stern–Gerlach Experiment

The theory of spatial quantization of the spin moment of the momentum of electrons of atoms situated in the magnetic field needed to be proved experimentally. In 1920 (two years before the theoretical description of the spin was created) Otto Stern and Walter Gerlach observed it in the experiment they conducted.

Silver atoms were evaporated using an electric furnace in a vacuum. Using thin slits, the atoms were guided into a flat beam and the beam sent through an in-homogeneous magnetic field before colliding with a metallic plate. The laws of classical physics predict that the collection of condensed silver atoms on the plate should form a thin solid line in the same shape as the original beam. However, the in-homogeneous magnetic field caused the beam to split in two separate directions, creating two lines on the metallic plate.

The phenomenon can be explained with the spatial quantization of the spin moment of momentum. In atoms the electrons are paired such that one spins upward and one downward, neutralizing the effect of their spin on the action of the atom as a whole. But in the valence shell of silver atoms, there is a single electron whose spin remains unbalanced.

The unbalanced spin creates spin magnetic moment, making the electron act like a very small magnet. As the atoms pass through the in-homogeneous magnetic field, the force moment in the magnetic field influences the electron's dipole until its position matches the direction of the stronger field. The atom would then be pulled toward or away from the stronger magnetic field a specific amount, depending on the value of the valence electron's spin. When the spin of the electron is $+1/2$ the atom moves away from the stronger field, and when the spin is $-1/2$ the atom moves toward it. Thus the beam of silver atoms is split while traveling through the in-homogeneous magnetic field, according to the spin of each atom's valence electron.

In 1927 Phipps and Taylor conducted a similar experiment, using atoms of hydrogen with similar results. Later scientists conducted experiments using other atoms that have only one electron in their valence shell: (copper, gold, sodium, potassium). Every time there were two lines formed on the metallic plate.

The atomic nucleus also may have spin, but protons and neutrons are much heavier than electrons (about 1836 times), and the magnetic dipole moment is inversely proportional to the mass. So the nuclear magnetic dipole momentum is much smaller than that of the whole atom. This small magnetic dipole was later measured by Stern, Frisch and Easterman.

Energy Levels from the Dirac Equation

In 1928, Paul Dirac developed a relativistic wave equation, now termed the Dirac equation, which predicted the spin magnetic moment correctly, and at the same time treated the electron as a point-like particle. Solving the Dirac equation for the energy levels of an electron in the hydrogen atom, all four quantum numbers including *s* occurred naturally and agreed well with experiment.

Total Spin of an Atom or Molecule

For some atoms the spins of several unpaired electrons (s_1, s_2, ...) are coupled to form a *total spin* quantum number S. This occurs especially in light atoms (or in molecules formed only of light atoms) when spin-orbit coupling is weak compared to the coupling between spins or the coupling between orbital angular momenta, a situation known as LS coupling because L and S are constants of motion. Here L is the total orbital angular momentum quantum number.

For atoms with a well-defined S, the multiplicity of a state is defined as (2S+1). This is equal to the number of different possible values of the total (orbital plus spin) angular momentum J for a given (L, S) combination, provided that $S \leq L$ (the typical case). For example, if S = 1, there are three states which form a triplet. The eigenvalues of S_z for these three states are $+1\hbar$, 0 and $-1\hbar$. The term symbol of an atomic state indicates its values of L, S, and J.

Magnetic Quantum Number

The magnetic quantum number (symbol m_l) is one of four quantum numbers in atomic physics. The set is: principal quantum number, azimuthal quantum number, magnetic quantum number, and spin quantum number. Together, they describe the unique quantum state of an electron. The magnetic quantum number distinguishes the orbitals available within a subshell, and is used to calculate the azimuthal component of the orientation of orbital in space. Electrons in a particular subshell (such as s, p, d, or f) are defined by values of ℓ (0, 1, 2, or 3). The value of m_l can range from $-\ell$ to $+\ell$, inclusive of zero. Thus the s, p, d, and f subshells contain 1, 3, 5, and 7 orbitals each, with values of m within the ranges 0, ±1, ±2, ±3 respectively. Each of these orbitals can accommodate up to two electrons (with opposite spins), forming the basis of the periodic table.

Derivation

These orbitals have magnetic quantum numbers $m = -\ell, \ldots, \ell$ from left to right in ascending order. The $e^{mi\phi}$ dependence of the azimuthal component can be seen as a color gradient repeating m times around the vertical axis.

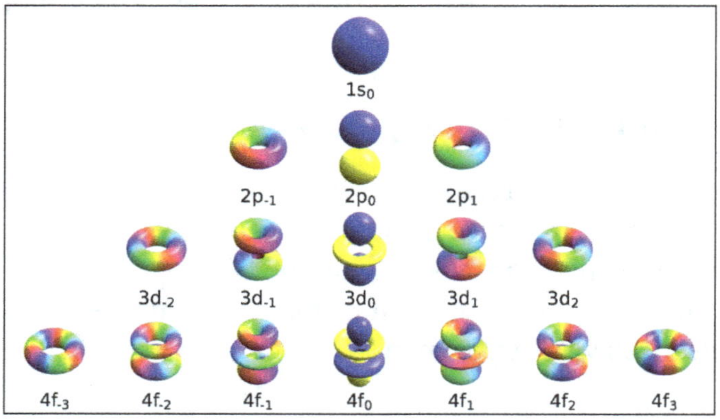

There is a set of quantum numbers associated with the energy states of the atom. The four quantum numbers n, ℓ, m_l, and s specify the complete and unique quantum state of a single electron in an atom called its wavefunction or orbital. The Schrödinger equation for the wavefunction of an atom with one electron is a separable partial differential equation. (This is not the case for the helium atom or other atoms with mutually interacting electrons, which require more sophisticated methods for solution). This means that the wavefunction as expressed in spherical coordinates can be broken down into the product of three functions of the radius, colatitude (or polar) angle, and azimuth:

$$\psi(r,\theta,\phi) = R(r)P(\theta)F(\phi)$$

The differential equation for F can be solved in the form $F(\phi) = Ae^{\lambda\phi}$. Because values of the azimuth angle ϕ differing by 2π (360 degrees in radians) represent the same position in space, and the overall magnitude of F does not grow with arbitrarily large ϕ as it would for a real exponent, the coefficient λ must be quantized to integer multiples of i, producing an imaginary exponent: $\lambda = im_l$. These integers are the magnetic quantum numbers. The same constant appears in the colatitude equation, where larger values of m_l^2 tend to decrease the magnitude of $P(\theta)$, and values of m_l greater than the azimuthal quantum number ℓ do not permit any solution for $P(\theta)$.

Relationship between Quantum Numbers			
Orbital	Values	Number of Values for	Electrons per subshell
s	$\ell = 0, \quad m_l = 0$	1	2

p	$\ell = 1, \quad m_l = -1, 0, +1$	3	6
d	$\ell = 2, \quad m_l = -2, -1, 0, +1, +2$	5	10
f	$\ell = 3, \quad m_l = -3, -2, -1, 0, +1, +2, +3$	7	14
g	$\ell = 4, \quad m_l = -4, -3, -2, -1, 0, +1, +2, +3, +4$	9	18

As a Component of Angular Momentum

Illustration of quantum mechanical orbital angular momentum. The cones and plane represent possible orientations of the angular momentum vector for $\ell = 2$ and $m = -2, -1, 0, 1, 2$. Even for the extreme values of m, the z component of this vector is less than its total magnitude.

The axis used for the polar coordinates in this analysis is chosen arbitrarily. The quantum number m refers to the projection of the angular momentum in this arbitrarily-chosen direction, conventionally called the z direction or quantization axis. L_z, the magnitude of the angular momentum in the z direction, is given by the formula:

$$L_z = m_l \hbar.$$

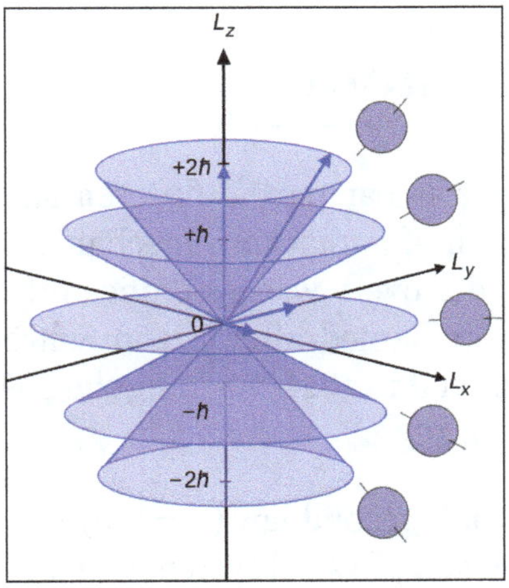

This is a component of the atomic electron's total orbital angular momentum, L, whose magnitude is related to the azimuthal quantum number of its subshell ℓ by the equation:

$$L = \hbar\sqrt{\ell(\ell+1)}$$

where $\hbar$ is the reduced Planck constant. Note this $L = 0$ for $\ell = 0$ and approximates $L = (\ell + 0.5)\hbar$ for high ℓ. It is not possible to measure the angular momentum of the electron in all three axes simultaneously. These properties were first demonstrated in the Stern-Gerlach experiment, by Otto Stern and Walther Gerlach.

The energy of any wave is its frequency multiplied by Planck's constant. The wave displays particle-like packets of energy called quanta. The formula for the quantum number of each quantum state uses Planck's reduced constant, which only allows particular or discrete or quantized energy levels.

Effect in Magnetic Fields

The quantum number *m* refers, loosely, to the direction of the angular momentum vector. The magnetic quantum number *m* only affects the electron's energy if it is in a magnetic field because in the absence of one, all spherical harmonics corresponding to the different arbitrary values of *m* are equivalent. The magnetic quantum number determines the energy shift of an atomic orbital due to an external magnetic field (the Zeeman effect) — hence the name *magnetic* quantum number. However, the actual magnetic dipole moment of an electron in an atomic orbital arrives not only from the electron angular momentum, but also from the electron spin, expressed in the spin quantum number.

Since each electron has a magnetic moment in a magnetic field, it will be subject to a torque which tends to make the vector *L* parallel to the field, a phenomenon known as Larmor precession.

Schrödinger Equation

The Schrödinger equation is a linear partial differential equation that describes the wave function or state function of a quantum-mechanical system. It is a key result in quantum mechanics, and its discovery was a significant landmark in the development of the subject. The equation is named after Erwin Schrödinger, who postulated the equation in 1925, and published it in 1926, forming the basis for the work that resulted in his Nobel Prize in Physics in 1933.

In classical mechanics, Newton's second law (F = ma) is used to make a mathematical prediction as to what path a given physical system will take over time following a set of known initial conditions. Solving this equation gives the position and the momentum of the physical system as a function of the external force F on the system. Those two parameters are sufficient to describe its state at each time instant. In quantum mechanics, the analogue of Newton's law is Schrödinger's equation.

The concept of a wave function is a fundamental postulate of quantum mechanics; the wave function defines the state of the system at each spatial position, and time. Using these postulates, Schrödinger's equation can be derived from the fact that the time-evolution operator must be unitary, and must therefore be generated by the exponential of a self-adjoint operator, which is the quantum Hamiltonian.

In the Copenhagen interpretation of quantum mechanics, the wave function is the most complete description that can be given of a physical system. Solutions to Schrödinger's equation describe not only molecular, atomic, and subatomic systems, but also macroscopic systems, possibly even the whole universe. Schrödinger's equation is central to all applications of quantum mechanics, including quantum field theory, which combines special relativity with quantum mechanics. Theories of quantum gravity, such as string theory, also do not modify Schrödinger's equation.

The Schrödinger equation is not the only way to study quantum mechanical systems and make predictions. The other formulations of quantum mechanics include matrix mechanics, introduced by Werner Heisenberg, and the path integral formulation, developed chiefly by Richard Feynman. Paul Dirac incorporated matrix mechanics and the Schrödinger equation into a single formulation.

Schrödinger's equation inscribed on the gravestone of Annemarie and Erwin Schrödinger. (Newton's dot notation for the time derivative is used).

Equation

Time-dependent Equation

The form of the Schrödinger equation depends on the physical situation. The most general form is the time-dependent Schrödinger equation (TDSE), which gives a description of a system evolving with time.

Time-dependent Schrödinger Equation (General)

$$i\hbar \frac{d}{dt} | \Psi(t) \rangle = \hat{H} | \Psi(t) \rangle$$

where i is the imaginary unit, $\hbar = \frac{h}{2\pi}$ is the reduced Planck constant, Ψ is the state vector of the quantum system, t is time, and $\hat{H}$ is the Hamiltonian operator. The *position-space wave function* of the quantum system is nothing but the components in the expansion of the state vector in terms of the position eigenvector $|\mathbf{r}\rangle$. It is a scalar function, expressed as $\Psi(\mathbf{r},t)=\langle\mathbf{r}\mid\Psi\rangle$. Similarly, the *momentum-space wave function* can be defined as $\tilde{\Psi}(\mathbf{p},t)=\langle\mathbf{p}\mid\Psi\rangle$, where $|\mathbf{p}\rangle$ is the momentum eigenvector.

The most famous example is the nonrelativistic Schrödinger equation for the wave function in position space $\Psi(\mathbf{r},t)$ of a single particle subject to a potential $V(\mathbf{r},t)$, such as that due to an electric field.

Time-dependent Schrödinger Equation in Position Basis (Single Nonrelativistic Particle)

$$i\hbar\frac{\partial}{\partial t}\Psi(\mathbf{r},t)=\left[\frac{-\hbar^2}{2m}\nabla^2+V(\mathbf{r},t)\right]\Psi(\mathbf{r},t)$$

where m is the particle's mass, and ∇^2 is the Laplacian.

This is also a diffusion equation, but unlike the heat equation, this one is also a wave equation given the imaginary unit present in the transient term.

The term *"Schrödinger equation"* can refer to both the general equation, or the specific nonrelativistic version. The general equation is indeed quite general, used throughout quantum mechanics, for everything from the Dirac equation to quantum field theory, by plugging in diverse expressions for the Hamiltonian. The specific nonrelativistic version is a strictly classical approximation to reality and yields accurate results in many situations, but only to a certain extent.

To apply the Schrödinger equation, write down the Hamiltonian for the system, accounting for the kinetic and potential energies of the particles constituting the system, then insert it into the Schrödinger equation. The resulting partial differential equation is solved for the wave function, which contains information about the system.

Time-independent Equation

The time-dependent Schrödinger equation described above predicts that wave functions can form standing waves, called stationary states. These states are particularly important as their individual study later simplifies the task of solving the time-dependent Schrödinger equation for *any* state. Stationary states can also be described by a simpler form of the Schrödinger equation, the *time-independent Schrödinger equation* (TISE).

Time-independent Schrödinger Equation (General)

$$\hat{\mathrm{H}}\mid\Psi\rangle=E\mid\Psi\rangle$$

where E is a constant equal to the energy level of the system. This is only used when the Hamiltonian itself is not dependent on time explicitly. However, even in this case the total wave function still has a time dependency.

In the language of linear algebra, this equation is an eigenvalue equation. Therefore, the wave function is an eigenfunction of the Hamiltonian operator with corresponding eigenvalues E.

As before, the most common manifestation is the nonrelativistic Schrödinger equation for a single particle moving in an electric field (but not a magnetic field):

Time-independent Schrödinger Equation (Single Nonrelativistic Particle)

$$\left[\frac{-\hbar^2}{2m}\nabla^2 + V(\mathbf{r})\right]\Psi(\mathbf{r}) = E\Psi(\mathbf{r})$$

with definitions as above. Here, the form of the Hamiltonian operator comes from classical mechanics, where the Hamiltonian *function* is the sum of the kinetic and potential energies. That is, $H = T + V = \frac{\|\mathbf{p}\|^2}{2m} + V(x, y, z)$ for a single particle in the non-relativistic limit.

Derivation

By assuming the Dirac-von Neumann axioms, a modern approach to understanding the Schrödinger equation is outlined as follows. Suppose the wave function $\psi(t_0)$ represents a unit vector defined on a complex Hilbert space at some initial time t_0. The unitarity principle requires that there must exist a linear operator, $\hat{U}(t)$, such that for any time $t > t_0$,

$$\psi(t) = \hat{U}(t)\psi(t_0).$$

Given that $\psi(t)$ must remain a unit vector, the operator $\hat{U}(t)$ must therefore be a unitary transformation. As such, there exists an exponential map such that $\hat{U}(t) = e^{-i\widehat{\mathcal{H}}t}$ where $\widehat{\mathcal{H}}$ is a Hermitian operator (given by the fact that the Lie algebra of the unitary group is generated by Hermitian operators). Therefore, the first-order Taylor expansion of $\hat{U}(t)$ centered at t_0 takes the form,

$$\hat{U}(t) \approx 1 - i(t - t_0)\widehat{\mathcal{H}} \cdot$$

And so, substituting the above expansion into $\psi(\mathrm{t}) = \hat{\mathrm{U}}(\mathrm{t})\psi(\mathrm{t}_0)$ thus yields,

$$\psi(t) = \psi(t_0) - i(t - t_0)\widehat{\mathcal{H}}\psi(t_0),$$

which, when rearranged and taken in the limit $t \to t_0$, provides an equation of the same form as the Schrödinger equation:

$$i\frac{d\psi}{dt} = \widehat{\mathcal{H}}\psi$$

where the ordinary definition for the derivative was used. However, the operator $\hat{\mathcal{H}}$ used here denotes an *arbitrary* Hermitian operator. Nonetheless, by using the correspondence principle it is possible to show that, in the classical limit and using appropriate units, the expectation value of $\hat{\mathcal{H}}$ indeed corresponds to the Hamiltonian of the system.

Properties

The Schrödinger equation has the following properties: some are useful, but there are shortcomings. Ultimately, these properties arise from the Hamiltonian used, and the solutions to the equation.

Linearity

In the development above, the Schrödinger equation was made to be linear for generality, though this has other implications. If two wave functions ψ_1 and ψ_2 are solutions, then so is any linear combination of the two:

$$\psi = a\psi_1 + b\psi_2$$

where *a* and *b* are any complex numbers (the sum can be extended for any number of wave functions). This property allows superpositions of quantum states to be solutions of the Schrödinger equation. Even more generally, it holds that a general solution to the Schrödinger equation can be found by taking a weighted sum over all single state solutions achievable. For example, consider a wave function $\Psi(x, t)$ such that the wave function is a product of two functions: one time independent, and one time dependent. If states of definite energy found using the time independent Schrödinger equation are given by $\psi_E(x)$ with amplitude A_n and time dependent phase factor is given by:

$$e^{-iE_nt/\hbar},$$

then a valid general solution is,

$$(x,t) = \sum_n A_n \psi_{E_n}(x) e^{-iE_nt/\hbar}.$$

Additionally, the ability to scale solutions allows one to solve for a wave function without normalizing it first. If one has a set of normalized solutions ψ_n, then,

$$\Psi = \sum_n A_n \psi_n$$

can be normalized by ensuring that,

$$\sum_n |A_n|^2 = 1.$$

This is much more convenient than having to verify that,

$$\int_{-\infty}^{\infty} |\Psi(x)|^2 dx = \int_{-\infty}^{\infty} \Psi(x)\Psi^*(x)dx = 1.$$

Momentum Space Schrödinger Equation

The Schrödinger equation $i\hbar\partial_t|\psi\rangle = \hat{H}|\psi\rangle$ is often presented in the position basis form $i\hbar\partial_t\psi(x) = -\frac{\hbar^2}{2m}\nabla^2\psi(x) + V(x)\psi(x)$ (with $\langle x|\psi\rangle \equiv \psi(x)$). But as a vector operator equation it has a valid representation in any arbitrary complete basis of kets in Hilbert space. For example, in the momentum space basis the equation reads,

$$i\ \partial\ f(p) = \text{——}\ f(p) + (V * f)(p)$$

where $|p\rangle$ istheplanewavestateofdefinitemomentum p, $\langle p|\psi\rangle \equiv f(p) = \int \psi(x)e^{-ipx}dx$, $\tilde{V}$ is the Fourier transform of V, and *denotes Fourier convolution.

In the 1D example with absence of a potential, $\tilde{V} = 0$ (or similarly $\tilde{V}(k) \propto \delta(k)$ in the case of a background potential constant throughout space), each stationary state of energy $\hbar\omega = q^2/2m$ is of the form:

$$f_q(p) = (c_+\delta(p-q) + c\_\delta(p+q))e^{-i\omega t}$$

for arbitrary complex coefficients $c_\pm$. Such a wave function, as expected in free space, is a superposition of plane waves moving right and left with momenta $\pm q$; upon momentum measurement the state would collapse to one of definite momentum $\pm qw$ with probability $\propto |c_\pm|^2$.

A version of the momentum space Schrödinger equation is often used in solid-state physics, as Bloch's theorem ensures the periodic crystal lattice potential couples $f(p)$ with $f(p+K)$ for only discrete reciprocal lattice vectors K. This makes it convenient to solve the momentum space Schrödinger equation at each point in the Brillouin zone independently of the other points in the Brillouin zone.

Real Energy Eigenstates

For the time-independent equation, an additional feature of linearity follows: if two wave functions ψ_1 and ψ_2 are solutions to the time-independent equation with the same energy *E*, then so is any linear combination:

$$\hat{H}(a\psi_1 + b\psi_2) = a\hat{H}\psi_1 + b\hat{H}\psi_2 = E(a\psi_1 + b\psi_2).$$

Two different solutions with the same energy are called *degenerate.*

In an arbitrary potential, if a wave function ψ solves the time-independent equation, so does its complex conjugate, denoted ψ^*. By taking linear combinations, the real and imaginary parts of ψ are each solutions. If there is no degeneracy they can only differ by a factor.

In the time-dependent equation, complex conjugate waves move in opposite directions. If $\Psi(x, t)$ is one solution, then so is $\Psi^*(x, -t)$. The symmetry of complex conjugation is called time-reversal symmetry.

Space and Time Derivatives

Continuity of the wave function and its first spatial derivative (in the *x* direction, *y* and *z* coordinates not shown), at some time *t*.

The Schrödinger equation is first order in time and second in space, which describes the time evolution of a quantum state (meaning it determines the future amplitude from the present).

Explicitly for one particle in 3-dimensional Cartesian coordinates – the equation is,

$$i\hbar\frac{\partial\Psi}{\partial t} = -\frac{\hbar^2}{2m}\left(\frac{\partial^2\Psi}{\partial x^2} + \frac{\partial^2\Psi}{\partial y^2} + \frac{\partial^2\Psi}{\partial z^2}\right) + V(x,y,z,t)\Psi.$$

The first time partial derivative implies the initial value (at $t = 0$) of the wave function,

$$\Psi(x,y,z,0)$$

is an arbitrary constant. Likewise – the second order derivatives with respect to space implies the wave function *and* its first order spatial derivatives,

$$\Psi(x_b,y_b,z_b,t)$$

$$\frac{\partial}{\partial x}\Psi(x_b,y_b,z_b,t) \quad \frac{\partial}{\partial y}\Psi(x_b,y_b,z_b,t) \quad \frac{\partial}{\partial z}\Psi(x_b,y_b,z_b,t)$$

are all arbitrary constants at a given set of points, where x_b, y_b, z_b are a set of points describing boundary *b* (derivatives are evaluated at the boundaries). Typically there are one or two boundaries, such as the step potential and particle in a box respectively.

As the first order derivatives are arbitrary, the wave function can be a continuously differentiable function of space, since at any boundary the gradient of the wave function can be matched.

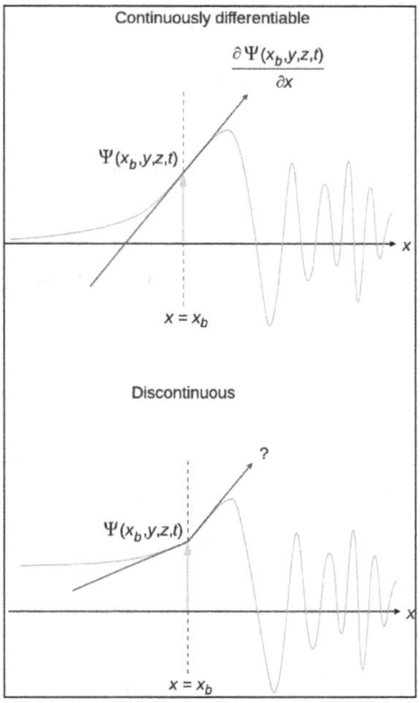

On the contrary, wave equations in physics are usually *second order in time*, notable are the family of classical wave equations and the quantum Klein–Gordon equation.

Local Conservation of Probability

The Schrödinger equation is consistent with probability conservation. Multiplying the Schrödinger equation on the right by the complex conjugate wave function, and multiplying the wave function to the left of the complex conjugate of the Schrödinger equation, and subtracting, gives the continuity equation for probability:

$$\frac{\partial}{\partial t}\rho(\mathbf{r},t)+\nabla\cdot\mathbf{j}=0,$$

where:

$$\rho=|\Psi|^2=\Psi^*(\mathbf{r},t)\Psi(\mathbf{r},t)$$

is the probability density (probability per unit volume, * denotes complex conjugate), and,

$$\mathrm{j}=\frac{1}{2m}\left(\Psi^*\hat{\mathrm{p}}\Psi-\Psi\hat{\mathrm{p}}\Psi^*\right)$$

is the probability current (flow per unit area).

Hence predictions from the Schrödinger equation do not violate probability conservation.

Positive Energy

If the potential is bounded from below, meaning there is a minimum value of potential energy, the eigenfunctions of the Schrödinger equation have energy which is also bounded from below. This can be seen most easily by using the variational principle, as follows.

For any linear operator $\hat{A}$ bounded from below, the eigenvector with the smallest eigenvalue is the vector ψ that minimizes the quantity,

$$\langle \psi | \hat{A} | \psi \rangle$$

over all ψ which are normalized. In this way, the smallest eigenvalue is expressed through the variational principle. For the Schrödinger Hamiltonian $\hat{H}$ bounded from below, the smallest eigenvalue is called the ground state energy. That energy is the minimum value of,

$$\langle \psi | \hat{H} | \psi \rangle = \int \psi^*(\mathbf{r}) \left[-\frac{\hbar^2}{2m} \nabla^2 \psi(\mathbf{r}) + V(\mathbf{r})\psi(\mathbf{r}) \right] d^3\mathbf{r}$$

$$= \int \left[\frac{\hbar^2}{2m} |\nabla \psi|^2 + V(\mathbf{r}) |\psi|^2 \right] d^3\mathbf{r} = \langle \hat{H} \rangle$$

(using integration by parts). Due to the complex modulus of ψ^2 (which is positive definite), the right hand side is always greater than the lowest value of $V(x)$. In particular, the ground state energy is positive when $V(x)$ is everywhere positive.

For potentials which are bounded below and are not infinite over a region, there is a ground state which minimizes the integral above. This lowest energy wave function is real and positive definite – meaning the wave function can increase and decrease, but is positive for all positions. It physically cannot be negative: if it were, smoothing out the bends at the sign change (to minimize the wave function) rapidly reduces the gradient contribution to the integral and hence the kinetic energy, while the potential energy changes linearly and less quickly. The kinetic and potential energy are both changing at different rates, so the total energy is not constant, which can't happen (conservation). The solutions are consistent with Schrödinger equation if this wave function is positive definite.

The lack of sign changes also shows that the ground state is nondegenerate, since if there were two ground states with common energy E, not proportional to each other, there would be a linear combination of the two that would also be a ground state resulting in a zero solution.

Analytic Continuation to Diffusion

The above properties (positive definiteness of energy) allow the analytic continuation of the Schrödinger equation to be identified as a stochastic process. This can be interpreted as the Huygens–Fresnel principle applied to De Broglie waves; the spreading wavefronts are diffusive probability amplitudes. For a free particle (not subject to a potential) in a random walk, substituting $\tau = it$ into the time-dependent Schrödinger equation gives:

$$\frac{\partial}{\partial \tau} X(\mathbf{r},\tau) = \frac{\hbar}{2m} \nabla^2 X(\mathbf{r},\tau), \quad X(\mathbf{r},\tau) = \Psi(\mathbf{r},\tau / i)$$

which has the same form as the diffusion equation, with diffusion coefficient $\frac{\hbar}{2m}$.

Regularity

On the space L^2 of square-integrable densities, the Schrödinger semigroup $e^{it\hat{H}}$ is a unitary evolution, and therefore surjective. The flows satisfy the Schrödinger equation $i\partial_t u = \hat{H}u$, where the derivative is taken in the distribution sense. However, since $\hat{H}$ for most physically reasonable Hamiltonians (e.g., the Laplace operator, possibly modified by a potential) is unbounded in L^2, this shows that the semigroup flows lack Sobolev regularity in general. Instead, solutions of the Schrödinger equation satisfy a Strichartz estimate.

Klopman-Salem Equation

In the theory of chemical reactivity, the Klopman-Salem equation describes the energetic change that occurs when two species approach each other in the course of a reaction and begin to interact, as their associated molecular orbitals begin to overlap with each other and atoms bearing partial charges begin to experience attractive or repulsive electrostatic forces. First described independently by Gilles Klopman and Lionel Salem in 1968, this relationship provides a mathematical basis for the key assumptions of frontier molecular orbital theory (i.e., theory of HOMO-LUMO interactions) and hard soft acid base (HSAB) theory. Conceptually, it highlights the importance of considering both electrostatic interactions and orbital interactions (and weighing the relative significance of each) when rationalizing the selectivity or reactivity of a chemical process.

Formulation and Interpretation

In modern form, the Klopman-Salem equation is commonly given as,

$$\Delta E = \Big(-\sum_{a,b}(q_a + q_b)\beta_{ab}S_{ab}\Big) + \Big(\sum_{k<\ell}\frac{Q_k Q_\ell}{\varepsilon R_{k\ell}}\Big) + \Big(\sum_{r}^{\text{occ.}}\sum_{s}^{\text{unocc.}} - \sum_{s}^{\text{occ.}}\sum_{r}^{\text{unocc.}} \frac{2(\sum_{a,b} c_{ra}c_{sb}\beta_{ab})^2}{E_r - E_s}\Big),$$

where:

- q_a is the electron population in atomic orbital a,
- β_{ab}, S_{ab} are the resonance and overlap integrals for the interaction of atomic orbitals a and b,
- Q_k is the total charge on atom k,
- ε is the local dielectric constant,
- $R_{k\ell}$ is the distance between the nuclei of atoms k and l,
- c_{ra} is the coefficient of atomic orbital a in molecular orbital r,
- and E_r is the energy of molecular orbital r.

Broadly speaking, the first term describes the closed-shell repulsion of the occupied molecular orbitals of the reactants (four-electron filled-filled interactions, *steric effects*). The second term describes the Coulombic attraction or repulsion between the atoms of the reactants (ionic contribution, *electrostatic effects*). Finally, the third term accounts for all possible interactions between the occupied and unoccupied molecular orbitals of the reactants (two-electron filled-unfilled interactions, *stereoelectronic effects*). Although conceptually useful, the Klopman-Salem equation seldom serves as the basis for energetic analysis in modern quantum chemical calculations.

Because of the difference in MO energies appearing in the denominator of the third term, energetically close orbitals make the biggest contribution. Hence, approximately speaking, analysis can often be simplified by considering only the highest occupied and lowest unoccupied molecular orbitals of the reactants (the HOMO-LUMO interaction in frontier molecular orbital theory). The relative contributions of the second (ionic) and third (covalent) terms play an important role in justifying hard soft acid base theory (HSAB), with hard-hard interactions governed by the ionic term and soft-soft interactions governed by the covalent term.

Born-Oppenheimer Approximation

The Born-Oppenheimer approximation is one of the basic concepts underlying the description of the quantum states of molecules. This approximation makes it possible to separate the motion of the nuclei and the motion of the electrons. This is not a new idea for us. We already made use of this approximation in the particle-in-a-box model when we explained the electronic absorption spectra of cyanine dyes without considering the motion of the nuclei. Then we discussed the translational, rotational and vibrational motion of the nuclei without including the motion of the electrons. In this chapter we will examine more closely the significance and consequences of this important approximation. Note, in this discussion nuclear refers to the atomic nuclei as parts of molecules not to the internal structure of the nucleus.

The Born-Oppenheimer approximation neglects the motion of the atomic nuclei when describing the electrons in a molecule. The physical basis for the Born-Oppenheimer approximation is the fact that the mass of an atomic nucleus in a molecule is much larger than the mass of an electron (more than 1000 times). Because of this difference, the nuclei move much more slowly than the electrons. In addition, due to their opposite charges, there is a mutual attractive force of $\frac{Ze^2}{r^2}$ acting on an atomic nucleus and an electron. This force causes both particles to be accelerated. Since the magnitude of the acceleration is inversely proportional to the mass, a = f/m, the acceleration of the electrons is large and the acceleration of the atomic nuclei is small; the difference is a factor of more than 1000. Consequently, the electrons are moving and responding to forces very quickly, and the nuclei are not. You can imagine running a 100-yard dash against someone whose acceleration is a 1000 times greater than yours. That person could literally run circles around you. So a good approximation is to describe the electronic states of a molecule by thinking that the nuclei aren't moving, i.e. that they are stationary. The nuclei, however, can be stationary at different positions so the electronic wavefunction can depend on the positions of the nuclei even though their motion is neglected.

Now we look at the mathematics to see what is done in solving the Schrödinger equation after making the Born-Oppenheimer approximation. For a diatomic molecule as an example, the Hamiltonian operator is grouped into three terms,

$$H(r,R) = \hat{T}_{nuc}(R) + \frac{e^2}{4\pi \in_0}\frac{Z_A Z_B}{R} + \hat{H}_{elec}(r,R)$$

Where:

$$T_{nuc}(R) = -\frac{\hbar^2}{2m_A}\nabla_A^2 - \frac{\hbar^2}{2m_B}\nabla_B^2$$

and,

$$\hat{H}_{elec}(r,R) = \frac{-\hbar^2}{2m}\sum_i \Delta_i^2 + \frac{e^2}{4\pi \in_0}\left(-\sum_i \frac{Z_A}{r_{Ai}} - \sum_i \frac{Z_B}{r_{Bi}} + \frac{1}{2}\sum_i \sum_{j\neq 1} \frac{1}{r_{ij}}\right)$$

In Equation $H(r,R) = \hat{T}_{nuc}(R) + \frac{e^2}{4\pi \in_0}\frac{Z_A Z_B}{R} + \hat{H}_{elec}(r,R)$, the first term represents the kinetic energy of the nuclei, the second term represents the Coulomb repulsion of the two nuclei, and the third term represents the contribution to the energy from the electrons, which consists of their kinetic energy, mutual repulsion for each other, and attraction for the nuclei. Bold-face type is used to represent that r and R are vectors specifying the positions of all the electrons and all the nuclei, respectively.

The Born-Oppenheimer approximation says that the nuclear kinetic energy terms in the complete Hamiltonian, Equation $H(r,R)=\hat{T}_{nuc}(R)+\frac{e^2}{4\pi\in_0}\frac{Z_AZ_B}{R}+\hat{H}_{elec}(r,R)$, can be neglected in solving for the electronic wavefunctions and energies. Consequently, the electronic wavefunction $\varphi_e(r,R)$ is found as a solution to the electronic Schrödinger equation:

$$\hat{H}_{elec}(r,R)\phi_e(r,R)=E_e(R)\phi_e(r,R)$$

Even though the nuclear kinetic energy terms are neglected, the Born-Oppenheimer approximation still takes into account the variation in the positions of the nuclei in determining the electronic energy and the resulting electronic wavefunction depends upon the nuclear positions, R.

As a result of the Born-Oppenheimer approximation, the molecular wavefunction can be written as a product the Born-Oppenheimer Approximation,

$$\psi_{ne}(r,R)=X_{ne}(R)\varphi_e(r,R)$$

This product wavefunction is called the Born-Oppenheimer wavefunction. The function $X_{ne}(R)$ is the vibrational wavefunction, which is a function of the nuclear coordinates R and depends upon both the vibrational and electronic quantum numbers or states, n and e, respectively. The electronic function, $\varphi_e(r,R)$, is a function of both the nuclear and electronic coordinates, but only depends upon the electronic quantum number or electronic state, e. Translational and rotational motion is not included here. The translational and rotational wavefunctions simply multiply the vibrational and electronic functions in Equation $\psi_{ne}(r,R)=X_{ne}(R)\varphi_e(r,R)$ to give he complete molecular wavefunction when the translational and rotational motions are not coupled to the vibrational and electronic motion.

Crude Born-Oppenheimer Approximation

In the *Crude Born-Oppenheimer Approximation*, R is set equal to R_0, the equilibrium separation of the nuclei, and the electronic wavefunctions are taken to be the same for all positions of the nuclei.

The electronic energy, $E_e(R)$, in Equation $\hat{H}_{elec}(r,R)\phi_e(r,R)=E_e(R)\phi_e(r,R)$ combines with the repulsive Coulomb energy of the two nuclei, to form the potential energy function that controls the nuclear motion as shown in figure.

$$V_e(R)=E_e(R)+\frac{e^2}{4\pi\epsilon_0}\frac{Z_AZ_B}{R}$$

Consequently the Schrödinger equation for the vibrational motion is:

$$(\hat{T}_{nuc}(R)+V(R))X_{ne}(R)=E_{ne\,Xne}(R)$$

The potential energy was approximated as a harmonic potential depending on the displacement, Q, of the nuclei from their equilibrium positions.

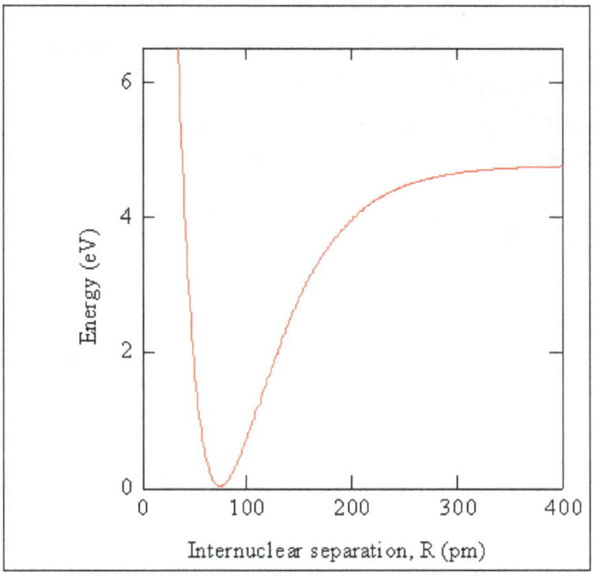

The potential energy function for a diatomic molecule.

In practice the electronic Schrödinger equation is solved using approximations at particular values of RR to obtain the wavefunctions $\varphi_e(r,R)$ and potential energies $V_e(R)$. The potential energies can be graphed as illustrated in figure.

The graph in figure is the energy of a diatomic molecule as a function of internuclear separation, which serves as the potential energy function for the nuclei. When R is very large there are two atoms that are weakly interacting. As R becomes smaller, the interaction becomes stronger, the energy becomes a large negative value, and we say a bond is formed between the atoms. At very small values of R, the internuclear repulsion is very large so the energy is large and positive. This energy function controls the motion of the nuclei. Previously, we approximated this function by a harmonic potential to obtain the description of vibrational motion in terms of the harmonic oscillator model. Other approximate functional forms could be used as well, e.g. the Morse potential. The equilibrium position of the nuclei is where this function is a minimum, i.e. at $R = R_0$. If we obtain the wavefunction at $R = R_0$, and use this function for all values of R, we have employed the *Crude Born-Oppenheimer* approximation.

References

- Atkins, Peter; de Paula, Julio; Friedman, Ronald (2009). Quanta, Matter, and Change: A Molecular Approach to Physical Chemistry. Oxford University Press. P. 106. ISBN 978-0-19-920606-3
- Quantum-Chemistry, Pr-Ro: chemistryexplained.com, Retrieved 21 May, 2020
- Shaik, Sason S.; Phillipe C. Hiberty (2008). A Chemist's Guide to Valence Bond Theory. New Jersey: Wiley-Interscience. ISBN 978-0-470-03735-5

- Licker, Mark, J. (2004). Mcgraw-Hill Concise Encyclopedia of Chemistry. New York: mc-graw-Hill. ISBN 978-0-07-143953-4.CS1
- Definition-of-quantum-number: thoughtco.com, Retrieved 8 January, 2020
- Halpern, Paul (2017-11-21). "Spin: The Quantum Property That Should Have Been Impossible". Forbes. Starts With A Bang. Retrieved 2018-03-10
- Andrew, A. V. (2006). "2. Schrödinger equation". Atomic spectroscopy. Introduction of theory to Hyperfine Structure. P. 274. ISBN 978-0-387-25573-6
- David Griffiths (2008). Introduction to elementary particles. Wiley-VCH. Pp. 162–. ISBN 978-3-527-40601-2. Retrieved 27 June 2011

PERMISSIONS

All chapters in this book are published with permission under the Creative Commons Attribution Share Alike License or equivalent. Every chapter published in this book has been scrutinized by our experts. Their significance has been extensively debated. The topics covered herein carry significant information for a comprehensive understanding. They may even be implemented as practical applications or may be referred to as a beginning point for further studies.

We would like to thank the editorial team for lending their expertise to make the book truly unique. They have played a crucial role in the development of this book. Without their invaluable contributions this book wouldn't have been possible. They have made vital efforts to compile up to date information on the varied aspects of this subject to make this book a valuable addition to the collection of many professionals and students.

This book was conceptualized with the vision of imparting up-to-date and integrated information in this field. To ensure the same, a matchless editorial board was set up. Every individual on the board went through rigorous rounds of assessment to prove their worth. After which they invested a large part of their time researching and compiling the most relevant data for our readers.

The editorial board has been involved in producing this book since its inception. They have spent rigorous hours researching and exploring the diverse topics which have resulted in the successful publishing of this book. They have passed on their knowledge of decades through this book. To expedite this challenging task, the publisher supported the team at every step. A small team of assistant editors was also appointed to further simplify the editing procedure and attain best results for the readers.

Apart from the editorial board, the designing team has also invested a significant amount of their time in understanding the subject and creating the most relevant covers. They scrutinized every image to scout for the most suitable representation of the subject and create an appropriate cover for the book.

The publishing team has been an ardent support to the editorial, designing and production team. Their endless efforts to recruit the best for this project, has resulted in the accomplishment of this book. They are a veteran in the field of academics and their pool of knowledge is as vast as their experience in printing. Their expertise and guidance has proved useful at every step. Their uncompromising quality standards have made this book an exceptional effort. Their encouragement from time to time has been an inspiration for everyone.

The publisher and the editorial board hope that this book will prove to be a valuable piece of knowledge for students, practitioners and scholars across the globe.

INDEX

A

Acetaldehyde, 15, 17, 53

Acetic Acid, 51, 53

Activation Energy, 7, 10, 27-32, 35, 38, 41, 48

Alkali Metals, 48, 86, 99

Amino Acid, 55, 67, 73, 80, 83

Ammonia, 39-40, 42, 44, 46, 48-50

Arrhenius Equation, 6-7, 10, 26-27, 29, 35

Atomic Mass, 2

B

Bauxite, 103, 137

Bimolecular Reaction, 10, 12, 24-25, 35, 59-60

Boltzmann Constant, 28, 30, 34, 177

C

Carbon Monoxide, 17, 42-43, 50, 53, 173

Catalysis, 1, 6, 15, 19, 37-40, 43-46, 48-55, 57-59, 64, 70-71, 75-76, 84

Cathode, 89, 104-105, 108, 110-111, 114-115, 117-118, 121, 125, 129-130, 132-134, 136-140

Chain Reaction, 10, 17, 38

Chemical Equilibrium, 6, 41, 83, 109, 182

Chemical Kinetics, 3, 6-9, 29, 32, 60, 84, 149

Chemical Thermodynamics, 1, 84, 142-145, 148, 178, 188

Chemiluminescence, 11

Chlorine, 17, 40, 48, 93, 100, 129, 132, 134, 136-140

Chymotrypsin, 64, 67

Collision Theory, 6, 27, 30-32, 34, 36-37

Corrosion Couple, 99

Crystal Lattice, 96-97, 225

D

Dihydrofolate Reductase, 54-55, 64, 73

E

Electrochemical Reaction, 85, 88, 90, 94-96, 104, 108, 111, 116, 118, 120, 188

Electrode Surface, 90-91, 96-97

Elementary Reaction, 12-13, 31, 34, 51, 59

Entropy, 3, 31, 109, 142-145, 147-148, 153-156, 159-163, 167, 169-170, 177-181, 183-184, 188, 190, 193

Enzyme Kinetics, 6, 54-55, 57-59, 62, 67, 71

Ethanol, 15, 37, 131

F

Ferrous Ions, 93, 130

Fluorescence, 57

Flux Balance Analysis, 25, 62

Free Radicals, 9, 17, 40

G

Galvanic Cells, 103, 109, 111, 135

Gibbs Energy, 31, 102, 143, 146, 167, 182, 186-189

Glucokinase, 72

H

Helmholtz Free Energy, 152, 184

Homogeneous Catalysis, 15, 40, 43, 51-54

Hydrogen Atom, 67, 195, 197, 206, 208-209, 211, 217

Hydrogenation Reaction, 36

I

Irreversible Inhibitors, 70, 73-74, 80-82

K

Kinetic Behaviour, 82

Kinetic Energy, 7, 30, 36, 157, 165, 176, 208, 228, 231-232

Kinetic Theory, 34

L

Lead-acid Battery, 101, 111, 122, 125, 127, 149

M

Mass Spectrometry, 11, 57, 81

Mercury, 92, 111, 135, 137-139, 141

Metabolic Pathways, 25, 62
Metal-solution Interface, 90, 92

N
Nickel, 42, 49-51, 100-101, 103, 118, 120, 125, 133
Nuclear Chemistry, 2
Nucleophilic Substitution, 10, 15-16

O
Oxidation State, 4, 85-86, 106

P
Peptide Bond, 64
Phosphofructokinase, 55, 65, 71
Photochemical Reactions, 3-4
Potential Energy, 9, 11, 101, 103, 145, 157-158, 165, 168, 197, 199, 228, 232-233

Q
Quantum Mechanics, 1, 4, 94, 194, 197-200, 202, 204-205, 207-208, 213-214, 220-222

R
Rate Constant, 12-14, 17, 25-32, 35, 59-60, 83
Rate Equation, 6, 9, 11-17, 19, 22, 24, 33, 35, 58, 81
Reaction Intermediates, 9-10, 13
Reaction Mechanism, 5-6, 8-12, 17, 57-58, 95
Redox Reaction, 4, 85-87, 92, 103, 149
Rhodium, 42-43
Ritonavir, 78-79

S
Sabatier Principle, 46-47
Sodium Hydroxide, 8, 132, 136-141
Sodium Metal, 8, 134
Spontaneous Process, 99, 142, 145, 148, 152-154, 190, 192
Stereochemistry, 9-10
Stoichiometric Coefficients, 12-13, 19, 25
Sulfuric Acid, 39, 42, 49, 111, 113
Surface Chemistry, 1

T
Thermal Energy, 3, 31, 155, 160, 172
Thermal Equilibrium, 3, 143, 151-152, 154-156, 178
Thermal Expansion, 1, 164, 170
Thermodynamic Cycle, 156, 162
Transition State, 8-9, 11, 13, 30, 41, 70-71, 79, 94

U
Unimolecular Reaction, 15, 26, 59

www.ingramcontent.com/pod-product-compliance
Lightning Source LLC
LaVergne TN
LVHW080620210926
848485LV00003B/191